Speculative Empiricism

Speculative Realism

Series Editor: Graham Harman

Books available

Onto-Cartography: An Ontology of Machines and Media, Levi R. Bryant
Form and Object: A Treatise on Things, Tristan Garcia, translated by Mark Allan Ohm and Jon Cogburn
Adventures in Transcendental Materialism: Dialogues with Contemporary Thinkers, Adrian Johnston
The End of Phenomenology: Metaphysics and the New Realism, Tom Sparrow
Fields of Sense: A New Realist Ontology, Markus Gabriel
Quentin Meillassoux: Philosophy in the Making Second Edition, Graham Harman
Assemblage Theory, Manuel DeLanda
Romantic Realities: Speculative Realism and British Romanticism, Evan Gottlieb
Garcian Meditations: The Dialectics of Persistence in Form and Object, Jon Cogburn
Speculative Realism and Science Fiction, Brian Willems
Speculative Empiricism: Revisiting Whitehead, Didier Debaise, translated by Tomas Weber

Forthcoming books

After Quietism: Analytic Philosophies of Immanence and the New Metaphysics, Jon Cogburn
Infrastructure, Graham Harman

Visit the Speculative Realism website at: edinburghuniversitypress.com/series/specr

Speculative Empiricism

Revisiting Whitehead

Didier Debaise

Translated by Tomas Weber

Edinburgh University Press is one of the leading university presses in the UK. We publish academic books and journals in our selected subject areas across the humanities and social sciences, combining cutting-edge scholarship with high editorial and production values to produce academic works of lasting importance. For more information visit our website: edinburghuniversitypress.com

Translated with the financial support of the University Foundation of Belgium

Edinburgh University Press Ltd
The Tun – Holyrood Road, 12(2f) Jackson's Entry, Edinburgh EH8 8PJ

Typeset in 11/13 Adobe Sabon by
Servis Filmsetting Ltd, Stockport, Cheshire,
and printed and bound in Great Britain by
CPI Group (UK) Ltd, Croydon CR0 4YY

A CIP record for this book is available from the British Library

ISBN 978 1 4744 2304 5 (hardback)
ISBN 978 1 4744 2306 9 (webready PDF)
ISBN 978 1 4744 2305 2 (paperback)
ISBN 978 1 4744 2307 6 (epub)

Contents

Series Editor's Preface

Although the great English philosopher Alfred North Whitehead (1861–1947) was born more than a century too early to be a Speculative Realist, he was without question speculative, and without question a realist. Hence there is no question that a well-executed book on Whitehead belongs in this series; Didier Debaise himself has expressed full agreement on this point. In the seventy years since his death, Whitehead has not quite secured the place in the Western philosophical canon that will one day surely be his. While philosophy for a century and more has been polarised between the Anglo-American analytic and Continental European traditions, Whitehead falls helplessly into a crack between the two. He is perhaps too speculative and wisemanish to impress the mechanical engineers of analytic philosophy who, if they admire Whitehead at all, do so only because of his mathematical work with their icon Bertrand Russell. This attitude is best symbolised by A. W. Moore's stunning omission of Whitehead, obviously a master metaphysician, from his book *The Evolution of Modern Metaphysics*: an otherwise warmly inclusive volume that reaches across the aisle to welcome such debonair continental figures as Bergson, Heidegger, Derrida, and even the snowball-throwing Deleuze.[1] Until recently, Whitehead was rarely read or mentioned by continentals either, being too different in kind from the first-wave German heroes (Kant through Heidegger) and second-wave French stars (primarily Derrida and Foucault) who filled up the pantheon of continental philosophy until the mid-1990s. Though some eccentric English and German scholars were reading Whitehead all along, for years his legacy was kept alive mainly

[1] A. W. Moore, *The Evolution of Modern Metaphysics: Making Sense of Things* (Cambridge: Cambridge University Press, 2012).

among American process theologians such as Charles Hartshorne (pronounced 'Heart's Horn') and John Cobb. Whitehead was as revered in these circles as he was ignored or chuckled over in the others.

The opening of new continental doors for Whitehead commenced with some positive references by Gilles Deleuze, who long made a habit of drawing our attention to important figures half- or fully forgotten: Albert Lautman, Gilbert Simondon, Étienne Souriau, Gabriel Tarde. In *The Fold*, his widely read book on Leibniz, Deleuze refers to Whitehead glowingly as Leibniz's rightful successor, and as 'provisionally . . . the last great Anglo-American philosopher before Wittgenstein's disciples spread their misty confusion, sufficiency, and terror'.[2] What continentally minded reader could resist such vindication, at the end of a century so unkind to continentals in institutional terms? Such was the prestige of Deleuze among those Francophone thinkers born just after the Second World War that most of his passing hints were eventually seized upon and developed by someone. In the case of Whitehead's gradual acquisition of continental citizenship, much of the credit goes to the Belgian philosopher Isabelle Stengers: severe and formidable in character, scientific in education, catholic in reading tastes. We are fortunate that Tomas Weber's beautiful English rendering of the present book is prefaced by Stengers herself: the one-time teacher of Debaise, her fellow Belgian. Like many others, I first became aware of Debaise's work through the recommendation of Bruno Latour, also an associate of Stengers – though of her own generation. It was in large part under Stengers's influence that Latour began to incorporate Whiteheadian ideas into his masterful reformation of philosophy and the social sciences. Latour made partial repayment for the favour with his Foreword to Stengers's own important book on Whitehead.[3]

Debaise the philosophical author has much in common with Debaise the person: formal, polite, soft-spoken, and also observant, tenacious and innovative. In her Preface below, Stengers is quick to note what is also one of my own favourite aspects of Debaise's book: his reminder that for Whitehead, 'philosophy is

[2] Gilles Deleuze, *The Fold: Leibniz and the Baroque*, trans. Tom Conley (London: Continuum, 2006), p. 86.

[3] Isabelle Stengers, *Thinking with Whitehead: A Free and Wild Creation of Concepts*, trans. Michael Chase (Cambridge, MA: Harvard University Press, 2014).

self-evident or it is not philosophy'.[4] With exquisite philosophical taste, what Debaise notices here is that beyond the intrinsic interest of any of Whitehead's philosophical propositions, he is one of the most insightful philosophers *about* philosophy in the Western tradition. Analytic philosophy has always been thoroughly rationalist in spirit, but in our time continental philosophy is riding an ascendant wave of rationalism, due in large part to the influence of Alain Badiou and his prominent student Quentin Meillassoux. After several decades of cloudy postmodernist weather, the spirit of mathematics, natural science and radical Enlightenment predominates once more. In this climate, philosophy aims at *proofs* in a manner reminiscent of Euclidean geometry. Whitehead, despite being a more accomplished mathematician than nearly any philosopher one could ever hope to meet, rejects this conception of what philosophy does. As he tells us in the opening pages of *Process and Reality*, philosophy does not proceed mainly through the rigorous logical deduction of eternally valid truths from unshakeable first principles – that confident method of great seventeenth-century thinkers like Descartes and Spinoza. Instead, philosophy's proper method is *descriptive generalization*: which aims not only at rigour, but also at not excluding anything that happens not to make a good fit with our initial prejudices. Ultimately, philosophy must also be aware of its inevitable failure to fathom the superfluous depths of the cosmos. Where analytic thinkers have an almost religious faith in the central role of argumentation in philosophy, Whitehead patiently cautions that philosophies are not refuted, but abandoned. That is to say, what ultimately kills off a philosophy is not some 'knockdown argument' (a phrase beloved among analytics), but a general and sometimes vague insufficiency in accounting adequately for everything we know to be included in the world. It is ironic that a similar point was made later by that rationalist darling, the philosopher of science Imre Lakatos.[5] As Lakatos teaches, there are never 'decisive experiments' in the history of science, because every scientific theory is born already falsified by counter-instances: for example, hundreds of unresolved problems in the gravitational theory of Newton before

[4] Alfred North Whitehead, *Modes of Thought* (New York: The Free Press, 1968), p. 49.

[5] Imre Lakatos, *The Methodology of Scientific Research Programs* (Cambridge: Cambridge University Press, 1978).

Einstein was even born. A scientific theory, like a philosophical one, is not a bundle of irrefutably rigorous arguments. Instead, it is a 'research programme' that might gradually come to be seen as either 'progressive' or 'degenerating', but not overthrown by a single strong counter-argument.

Another powerful aspect of Debaise's book is his unusual awareness of the decisions guiding his own particular reading of Whitehead. Above all, he pays close attention to Whitehead's concept of 'societies', as distinguished from the 'actual entities' that are often seen as analogous to the Leibnizian monads. The central claim of his book, Debaise reports, is that 'experience is defined by the particular relations between . . . what Whitehead calls "societies". The concept of "societies", then, is given a central place, in so far as it crystallises the relations between method and existence. The speculative project locates the key to its relation with experience through this concept.' Debaise is fully aware of Whitehead's rather unorthodox use of the word 'society', which is not exclusively human and sometimes not human at all, as when molecules or stars form societies of their own. And here we find ourselves outside the orbit of mainstream post-Kantian philosophy, which is so narrowly concerned with relations involving humans and world, and so convinced that nothing can be said about nonhuman–nonhuman relations that is not already said better by the natural sciences. Debaise is too natural a speculative thinker to give in to such artificial and premature restrictions on the subject matter of philosophy. Indeed, he is perfectly suited as an interpreter of Whitehead, whom he reads not just accurately, but beautifully. I have long been puzzled by the lack of an English translation of this highly regarded book, and am proud that this series has made such a translation possible.

Graham Harman
Dubuque, Iowa
January 2017

Translator's Note

In the first section of what follows Didier Debaise points to Alfred North Whitehead's interest in philosophical points of no return. In *Process and Reality* Whitehead writes that philosophy 'never reverts to its old position after the shock of a great philosopher',[1] and in an earlier work, *Science and the Modern World*, he affirms Descartes's *Discourse on Method* and William James's essay 'Does Consciousness Exist?' as examples of such moments.[2] This is not because conceptions of reality contained in these works accurately represent some sort of truth about the world. It is not even thanks to any supposed power in the visions of the world these texts hold. It is because they craft new problems, or at least transform the stakes of old ones.

The present book argues that if Whitehead deserves to join Descartes and James (along with many others) at the table of 'shocking' philosophers it is because he effects a transformation on the set of problems characterising empiricism. Whitehead pushes empiricism from philosophical anthropology, an approach in which what matters is the subject's encounter with given data, to an encounter with the problem of existence as it is in itself. But this detour into metaphysics is not performed just for fun – it also responds to a conceptual necessity. The tendency of much philosophy to reduce the diversity of experience to a pre-given single element that could serve as its foundation is, for Whitehead, ultimately incoherent. If experience is to be elucidated, then, in a way that affirms its irreducibility to, say, a subject considered as its ground, then speculative metaphysics has to be transformed

1 Whitehead, *Process and Reality*, p. 11. Subsequent references to *Process and Reality* will be indicated by PR in the main text.

2 See Whitehead, *Science and the Modern World*, p. 143.

into a method not for affirming what the world *really* is, as if from nowhere, but rather for *interpreting*, making sense of, ungrounded multiplicities of experience.

New problems and questions arise from this amalgamation of speculative philosophy and empiricism, and Debaise outlines how *Process and Reality* goes about clarifying and responding to them. Can experience be thought systematically without transforming its richness into reductive abstractions? Can the existence of everything that exists be accounted for without losing the particularity of our experience, everything 'enjoyed, perceived, willed or thought' (PR, 3)? Can the methods of empiricism ever be reconciled with modes of systematic cosmological speculation?

Debaise's reading of *Process and Reality* dramatises what a philosophy that answers 'yes' to these questions looks like, how it works, what its stakes are. And the result is indeed shocking: a kind of philosophy that would 'exclude nothing',[3] an 'open adventure'[4] in which the limits of thought and the edges of the world's objects – its 'societies' – have not been established in advance but are gradually formed through affective and experiential processes and relations: prehensions. With speculative empiricism it becomes *not only possible* but *self-evidently necessary* to sever the exclusive links between experience and consciousness as well as to think experience as involved in the universe as a whole. The limits of thought, then, are not real limits: they are wide open.

I have used Whitehead's original technical terminology wherever possible: 'togetherness' for *être-ensemble*, 'environment' for *milieu*, 'route' for *trajectoire*, 'causal efficacy', when appropriate, for *causalité efficiente*; 'presentational immediacy' for *présentation immédiate*, 'the many', when appropriate, for *diversité* or *pluralité*, 'nexūs' as the plural of 'nexus'. Any failures to communicate the intended meaning of the original French are my own responsibility.

I thank the following for helping this project come to fruition, supporting me throughout the process and patiently responding to my questions: Didier Debaise, John DeWitt, Ersev Ersoy, Graham Harman, Hannah Kirklin, Carol Macdonald, Matt Phillips and Joshua Richeson.

[3] Whitehead, *Modes of Thought*, p. 2.

[4] See below, p. 111.

Preface

Isabelle Stengers

Most of those who have succeeded in plotting a path through *Process and Reality* have done so in the form of an 'introduction', or, one might say, an 'education' in the sense given to the term by Foucault in *The Hermeneutics of the Subject.* Foucault connects 'education' not to *educare* but to *educere*: an outstretched hand, leading us towards an outside. Outside of what, though, in this case? Perhaps, first of all, outside of the habits formed at university desks at which we learn to 'read a philosophy', yes, as it asks to be read, but in a mode in which our respect excludes our making it accountable for the manner in which it deals with what matters to us.

It is said that the first philosophy lecture Whitehead ever attended was his own, a mathematician turned, at the end of his life, philosopher. If he was committed to philosophy, then, it was because the questions posed there mattered to him; it had nothing to do with choosing a subject out of the range offered by the university. Whitehead had read the 'great philosophers' long before becoming a philosopher, of course. But he read them in a way that 'took them seriously', that is, without first submitting to their greatness. He asked the question of what they do to us, what they ask us to forget.

What the 'great philosophers' – and particularly the modern philosophers (Descartes, Locke, Hume, Kant) – ask us to forget finds its clearest expression in his final book, *Modes of Thought*: '[t]he question What *do* we know?, has been transformed into the question, What *can* we know?'[1] Aiming to restore the idea that, whatever the philosophers say, it *matters* what we actually know, he cannot behave like a respectful philosopher in the face of the

[1] Whitehead, *Modes of Thought*, p. 74.

history of philosophy's habits. The task commits him, rather, to going hand-to-hand with earlier philosophers.

Didier Debaise's decision, then, to organise a reading of *Process and Reality* around the question of experience is extremely significant. With modern philosophy, for Whitehead, (conscious) experience becomes the source of primary data, what he calls the 'subjectivist principle'. Whitehead responds to this principle by 'reforming' it: it is clear that we should never admit that which cannot be discerned as an element of experience into a philosophical scheme (PR, 166), and yet 'apart from the experiences of subjects there is nothing, nothing, nothing, bare nothingness' (PR, 167). Not the nothingness invoked by phenomenologists in order to bypass Kant's 'things in themselves' but a nothingness that will become exposed to a vertiginous adventure in which *the category of the subject is expanded to include everything that we know exists outside of the experience of the human subject.* The Whiteheadian concept of experience, then, forces us to leave solid ground behind to follow what Deleuze and Guattari called a 'witch's line'. Everything that the subjectivist principle attempted to bracket in the name of what we *can* know returns, stripped of anything that might have assured its 'epistemological' domestication.

For Whitehead, however, experience is not just a concept. It is also what is at stake. At the beginning of this book, Debaise announces one of the strangest aspects of Whitehead's proposal. While the concepts expressed by his philosophical proposition might seem, at first glance, to pose a challenge to conceptions of philosophy based on 'reasonableness' (yes, but nevertheless, we can't say that. . .), Whitehead affirms, peacefully, that 'either philosophy is self-evident or it is not philosophy'.[2]

The paradox can be clarified if we take seriously what follows: 'the attempt of any philosophic discourse should be to produce self-evidence'.[3] Produce, not report. Above all, not to report on what is *given* as self-evident, 'I see an apple on this table.' The goal is to produce self-evidence in such a way that a report of a given transparent or consensual experience seems a masterpiece of sophistication, as remarkable for what it fails to disclose as for what it produces.

[2] Whitehead, *Modes of Thought*, p. 49.
[3] Whitehead, *Modes of Thought*, p. 49.

The question of experience, then, takes on a non-conceptual meaning. It concerns the experience of the reader, and the manner in which Whitehead's concepts, such as experience, alter experience itself, producing a double self-evidence: the self-evidence of the limited character of what we consider ourselves entitled to know in relation to what we *actually* know, as well as the self-evidence of the possibility of resisting this mode of judgement. In a famous metaphor Whitehead speaks of the flight of an aeroplane: 'and it again lands for renewed observation rendered acute by rational interpretation' (PR, 5). This is not just a metaphor. The idea should be taken literally: conscious observation, inhabited by the contrast between what is observed and the way in which what is observed might be interpreted, requires equipment, just like a laboratory. Everyday language is such a piece of equipment: its specialised abstractions are entirely appropriate for the everyday stakes of human life but are impressively deficient at meeting the challenge of speculative philosophy in Whitehead's sense, the challenge of becoming capable of making everything that we know matter to us.

Debaise, then, is absolutely right to begin with the question of speculative thought, characterised as the object of *Process and Reality*, and to link speculation to method. Whitehead's innocence is put on. His speculative thought is no return to any pre-Kantian 'great rationalism'. It is instead a comic mime, coherent with its thesis according to which 'philosophy never reverts to its old position after the shock of a great philosopher' (PR, 11). Whitehead knows very well that modifying experience in this way requires grappling with the philosophers behind the construction of the subjectivist principle who, in its name, undertook to purify experience of that which is considered not to constitute a primary given for philosophy. He also knows that the matter seems finished: speculative thought has been condemned, it belongs to a past, a page all these philosophers agree has been turned.

Perhaps the page has been turned. Nothing, after all, transcends experience. But isn't purified experience itself governed by a call to some sort of transcendence, a plea for givens that could be qualified as 'primary', purified of all interpretation? Whitehead responds to this search for 'pure' givens with a pragmatic method in William James's sense, a method involving the conscious, processual, experimental exploration of that which is capable of testifying to immediate experience. It is an immediacy, however, that can never be stripped of interpretation. 'If we desire a record of uninterpreted experience, we must ask a stone to record its autobiography' (PR, 15).

The flight of an aeroplane, the experimentation with the relation between concepts and immediate experience and with what experience, once renewed, demands from concepts, is inextricable from what Whitehead defines as ultimate: creativity. Whitehead, however, avoids the trap of 'intellectual intuition' within which we might expect to find speculative philosophers after Kant. There is no intuition of creativity as such since the latter is actual only by virtue of 'its accidents', that is, its actualisations, none of which have a privileged relation with it. Each intuition, from the most limited to the most sweeping, is an actualisation of creativity. What distinguishes the experiences activated by Whitehead's method is not that they 'attain creativity' as an ultimate principle of everything, but that they affirm themselves as creatures of creativity, productions of new existences and not as modes of knowledge finally adequate to a 'reality', whatever the philosophical meaning attached to that term.

Debaise, then, speaks of the 'considerable difference' between two points of view. According to the first point of view, Whitehead's 'system' would articulate general categories in a mode the ambition of which would be to account for experience in a coherent, necessary and logical manner. From the second point of view, however, coherence, logic and necessity are understood as 'methodical constraints', as definitions of what has to satisfy rational interpretation. From this second perspective, then, which Debaise takes up, Whitehead's thought becomes doubly speculative. It is speculative in the classical sense, of course, since Whitehead's categories are indeed unfolded in a conceptual space that is supposed to be closed off to philosophers. But it is also speculative in the ordinary sense of the word, in the sense of speculating about a possibility. Whitehead wagers that his system, a system that flaunts its affiliations with the rationalist tradition, that submits to constraints that could be considered rational demands, is capable of exposing this tradition to adventure. He wagers that it is capable of stripping the tradition of its powers of classification or exclusion.

Debaise calls Whitehead's wager 'rationalist trust'. When Whitehead proposes to 'rescue' the style of thought of Bergson, James and Dewey 'from the charge of anti-intellectualism, which rightly or wrongly has been associated with it' (PR, xii), it should be clear that it hardly matters whether such a charge is justified. What counts is simply the fact that it might be taken in that way, that is, that readers of Bergson, James or Dewey might understand 'rationalist' abstraction as in some way separate from 'reality'. Whitehead, however, speculates on the possibility of using abstractions to modify experience, abstractions produced through a search for 'reasons' and through an explicit demonstration of their mode of production. Not so that experience can then finally become 'rational', but rather to experiment with the possibility of expressing a disconnect between reason and submission by taking the question of reason as far as it is possible to go. For Whitehead, no definition is effective without the mute appeal to making an imaginative leap (PR, 4). Whitehead's reasons have as their mission to make this leap felt.

This second point of view demands method never be overlooked, as when admiring a building we might forget the scaffolding after its removal. When we read: '[t]he universe is at once the multiplicity of *res verae* and the solidarity of *res verae*' (PR, 167), and 'the concrescence of a *res vera* is the development of a subjective aim' (PR, 167), it is not about knowing if such a 'vision' can be accepted. What Whitehead calls *res verae*, actual entities connecting existence and actualisation in a mode which discloses them as creatures of creativity, have nothing to do with the kind of truth that would require contemplation or compliance. They are 'true' in the pragmatic sense, connecting truth to an experimentation with consequences. To understand them is, first of all, to understand the problem to which they respond, the manner in which they satisfy the method's constraints.

Debaise's essay contains two singularities, and they both find their origin here. On the one hand, each aspect of Whitehead's scheme is understood as an object of a problematic approach: the concept is considered through Whitehead's grappling with the philosophical tradition, itself subjected to methodical constraints. On the other hand, the greatest attention is devoted to the distinction between the mode of experience associated with actual entities and our mode of experiencing, the mode the method proposes to expose to adventure.

These two singularities are connected, and the power of Debaise's approach lies in the way it makes clear their inseparable character. Debaise casts light on the overwhelming importance, for Whitehead, of challenging the false simplicity of perceptual data, the pitfall of the empiricist tradition. This determined and relentless challenge to the empiricist reduction of experience has led many to misread Whitehead's construction as a phenomenological project. But Debaise emphasises that the stakes of his fight with Hume, in particular, are to be found on the level of conceptual construction and have little to do with fidelity to human experience.

In fact, this is why Whitehead is quick to offer arguments likely to bewilder phenomenologists just as much as empiricists. On the topic of Hume's professed ignorance regarding the 'causes' of visual sensations in our souls, Whitehead, with his comic use of common sense, introduces the difference between a disturbance in vision linked to a defect in the sharpness of vision or to a visual disturbance connected to alcohol: '[i]f the causes be indeed unknown, it is absurd to bother about eye-sight and intoxication. The reason for the existence of oculists and prohibitionists is that various causes *are* known' (PR, 171). It is not, for Whitehead, about constructing 'causes' of perception, or perceptual disturbances. Hume pretends to be unaware not of Whitehead's concepts but of the knowledge of oculists and drunk drivers. It is about what 'we know', what Hume has set about making us forget.

Whitehead's implacable reading of Hume and Locke, shedding light on their wilful ignorance and their sleights of hand, is not intended to offer a phenomenology of perception free of such defects but is part of a double movement that continues across *Process and Reality*. Whitehead attempts to make felt, by pointing to a diversity of causes, the speciousness of taking the visual relation as a simple point of departure, adequate for conceptual construction. On the other hand, however, he mobilises empirical philosophers who have pushed the modern 'subjectivist principle' as far as it will go, who have erased all reasons that transcend experience with the aim of 'reforming' the principle, effecting the dizzying passage from human experience to the experience of every actualisation of creativity.

In his own philosophy, Whitehead writes, '[i]n the philosophy of organism "the soul" as it appears in Hume, and "the mind" as it appears in Locke and Hume, are replaced by the phrases "the actual entity", and "the actual occasion", these phrases being synonymous' (PR, 141). What we know about vision, that there are causes, indeed has to take on an irreducible meaning concerning the experiences of actual entities or occasions. If Whitehead's scheme were to succeed in denying, or in returning to a quasi-invincible illusion, that which the practices of oculists and the protestations against 'drunk driving' both require, namely, that there are causes of vision, the scheme would have yielded to the facility Whitehead wishes to refuse to philosophy: the 'prowess' of explaining a meaningful aspect of experience by getting rid of it. In contrast, the interpretation of what is known by oculists and prohibitionists is part of the scheme, independent of the specialised languages used by each.

Debaise, then, proposes a radical distinction between the methodical construction proper to the speculative thinking of experience, in the sense of the 'reformed subjectivist principle', and the experience to which we tend to bear witness – perceptual experience of the kind testified to by empiricists. And this distinction is connected to another invaluable singularity of his essay. The Whiteheadian concept of 'society' is too often underestimated by his readers, reduced to a few brief examples casting light on the difference between, say, electrons and living cells. And yet it could be claimed that, independent of the societies to which they belong, the Whiteheadian *res vera* is a *dangerous abstraction* for the fact that it provokes a short circuit between its conceptually constructed relation with creativity, the ultimate, and the human experiences we call 'creative'. The two are not without relation, of course, but the relation has nothing obvious about it. The role of Whiteheadian societies is to produce this mediation, that is, to provide the environment that 'our' creative experiences require, while at the same time avoiding any dualism between 'habit' and creativity.

The third part of what follows, devoted to Whiteheadian societies, is therefore at once necessary, a direct consequence of the role given to method, and original, opening *Process and Reality* to questions that Whitehead calls for but does not himself take up, the value of his scheme being only pragmatic, shaped by the 'applications' it provokes. Didier Debaise introduces Whitehead's work, then, but he also adds to it by producing a path to consequences that solidify Whitehead's place in the pragmatic tradition.

To 'introduce', where Whitehead is concerned, becomes identified with 'educating' in the sense of 'holding out a hand'. It means to 'lead outside', yes, but it also involves making the two gestures of an outstretched hand inseparable: the leading away from reasonable and plausible reassurances, of course, but also the offering of support to those threatened by vertigo, those who have taken the path of Whitehead's departure. Such support, however, in Whitehead's case, must itself by taken in the midst of a creative process. Outstretched hands must not reassure but must rather provoke the experience of the 'aeroplane' without which the work would reach a dead-end. As Deleuze and Guattari write about the brain in *What is Philosophy?*,[4] we finish not by thinking recognition but by 'penetrating' a text shaped by an open series of imaginative leaps, a movement that is in itself *creative*.

Université Libre de Bruxelles

[4] Deleuze and Guattari, *What is Philosophy?*, pp. 209–10.

1

Introduction

Every reading of *Process and Reality* has to start by declaring the initial problem on the basis of which the book is to be interpreted. It has become clear to readers of Whitehead that different perspectives never stop reorganising the system, perspectives that determine relative and changing areas of importance, connecting problems with ever fluctuating forms. One of *Process and Reality*'s particularities – connected to its style and philosophical form – is that it resists being 'surveyed' from above, it resists readings introduced as mere explanations. As such, 'there are distinct lineages of readers of Whitehead in accordance with different approaches. Each reader inherits one particular movement out of all the movements that the original work tangles up.'[1] These tangles are the transformative power of the system to which one must become sensitive. An adequate reading cannot avoid taking a particular path.[2] The only constraints are those shared by all systematic thought: coherence must be maintained.

One could say of the interpretations of Whitehead's work what Whitehead himself considers a requirement for all philosophical thought:

> philosophy, in any proper sense of the term, cannot be proved. For proof is based upon abstraction. Philosophy is either self-evident, or it is not philosophy. The attempt of any philosophic discourse should be to produce self-evidence.[3]

Proof matters little. What does matter, however, is the capacity to *produce self-evidence*. It is this capacity that prevents all relativism concerning the success and relevance of a reading. One can arrive at a point of self-evidence from the technical and systematic constraints that characterise what *Process and Reality* is, not in its entirety but as one of the singular possibilities that it includes. The habitual order of explanation and demonstration, then, is reversed. It is generally assumed that if something is self-evident then it has no need of explanation, and so can function as an effective point of departure for the justification of a thought. But this is to forget everything that self-evidence involves, everything it implicitly recovers and extends. If ideas, habits or thoughts are self-evident, this is because they are the ends, not the causes, of a process. Self-evidence, then, should never be the initial element of a philosophical explanation but rather its goal. This presents a much greater challenge.

This reading of *Process and Reality* is organised around the concept of 'experience'. This is not an exclusive choice, but it does require taking a particular path through the work. It is not that the question of experience is absent from the work – far from it – but the importance and insistence that I want it to carry necessarily alters the system's economy. At the end of this reading, experience must have become self-evident. Rather than being an initial moment, a point of origin from which the system could be unfolded as the intensification of those initial impulses provided by experience, it becomes, rather, the object of the construction. There are two conditions for success, then, conditions that are at the origin of what I will term a 'speculative empiricism': to maintain the coherence of ideas, and to produce the self-evidence of experience. This, then, amounts to a radical empiricism since the ambition, here, is to elucidate immediate experience in its pluralism. When the concept of experience shifts, however, when it is transported from the beginning to the end of the elucidation, empiricism is transformed into a form of speculative thought, conceptual and abstract. The concept of experience, then, becomes inextricable from a form of philosophy that is the overall object of this reading.

The reading is organised into three parts that can be understood as the three components of such a speculative empiricism: 1) speculative philosophy is essentially a method; 2) this method makes it possible to define existence on the basis of creativity; 3) experience is defined by the particular relations between such existences, what Whitehead calls 'societies'. The concept of 'societies', then, is given a central place in so far as it crystallises the relations between method and existence. This concept is the key to the speculative project's relation with experience. Elucidation, then, becomes the invention of a space at once empirical and abstract (this distinction disappears at the end of the speculative movement) the goal of which is to express what it could mean to 'have' an experience without limiting experience by any single definition.

Notes

1. Stengers, *Penser avec Whitehead*, p. 18. Translator's note: This is my own translation, since this sentence has been excluded from the English edition (see Stengers, *Thinking with Whitehead*).
2. *Process and Reality* was published in 1929, following some of the most important events that have given birth to contemporary philosophy: Bergson's *Creative Evolution* (1907), James's *A Pluralistic Universe* (1909), Husserl's *Ideas I* (1913), Heidegger's *Being and Time* (1927) and Carnap's *The Logical Structure of the World* (1928). Each of these events, through their inheritances in French and American philosophy, have produced specific readings of *Process and Reality*. One of the first studies of Whitehead in French was that of Jean Wahl, who quite clearly chose a Heideggarian, and occasionally Bergsonian, approach (Wahl, *Vers le concret*). More recently, Bertrand Saint-Sernin has favoured a reading close to phenomenology in the Husserlian sense (Saint-Sernin, *Whitehead, un univers en essai*). Among these readings, despite key differences in orientation, a Deleuzian tradition can be noted, formed around *Difference and Repetition* and *The Fold*: see, for instance, Stengers, *Thinking with Whitehead*, and Dumoncel, *Les sept mots de Whitehead*.
3. Whitehead, *Modes of Thought*, p. 49.

Part I

Speculative Philosophy: Method and Function

2

What is Speculative Philosophy?

The Invention of a Method

A powerful proposition runs throughout *Process and Reality*: *speculative philosophy is essentially a method*. This proposition, despite its centrality, is announced explicitly only once, in the opening sentences of the book, at the start of the chapter entitled 'Speculative Philosophy'.

> This course of lectures is designed as an essay in Speculative Philosophy. Its first task must be to define 'speculative philosophy', and to defend it as a method productive of important knowledge. (PR, 3)

The French translators of *Process and Reality* chose to translate 'important knowledge' as 'wide-ranging knowledge' [*savoir de grande portée*].[1] This choice does not seem justified, even if it does respond to a legitimate concern about bringing to light particular aspects of knowledge. It overlooks the fact that the term 'importance' is a technical concept in *Process and Reality* and that Whitehead dedicates an entire chapter to it in *Modes of Thought*. 'Important knowledge', of course, is knowledge possessed with extension, and so might indeed be wide-ranging. This, however, is a consequence of its importance and not its essential characteristic.

In this chapter, speculative thought appears not as a merely introductory or preparatory question but is identified as the very object of the book. Whitehead announces this clearly by his introducing *Process and Reality* as an 'essay in Speculative Philosophy'. The stakes, then, are clear at first sight, and it is in the *literality* of what he says that the meaning and reason can be found for the task that takes him to the heart of a group of questions traditionally reserved for metaphysics.

Immediately, defining speculative thought amounts to organising, displaying and constructing a *method*. By making this question the very object of the book, Whitehead hardly ignores the fact that he is placing himself in a complex history loaded with implications and prejudices. In fact, it might seem surprising that *Process and Reality*'s project is at this point identified with the question of method: the book's subtitle is, after all, 'an essay in cosmology', and Whitehead often presents his project in metaphysical terms. How can all these elements be reconciled? How to move from an announcement rooted in cosmological or metaphysical language to a question the ambition of which, at this point, seems significantly diminished in comparison: that of method? The only possible hypothesis, if these questions are not to be sidelined, is that the term 'method' has undergone a shift in meaning.

Whitehead had already posed this question in *Science and the Modern World*, an earlier work than *Process and Reality*. There, he situated it historically in the wake of William James's pragmatism:

> in attributing to Williams James the inauguration of a new stage in philosophy, we should be neglecting other influences of his time. But, admitting this, there still remains a certain fitness in contrasting his essay, *Does Consciousness Exist*, published in 1904, with Descartes' *Discourse on Method*, published in 1637. James clears the stage of the old paraphernalia; or rather he entirely alters its lighting.[2]

The essay 'Does Consciousness Exist?' demands as many redefinitions, transformations and displacements for contemporary thought as the *Discourse on Method* did for the modern age. The comparison of the two works is important because it assumes a point of convergence, despite the fact that they radically differ in the way they define and set up the problem. James's invention of a new economy of problems should be understood in this way, as an original relation between elements of experience. When Whitehead writes, in *Process and Reality* this time, that '[p]hilosophy never reverts to its old position after the shock of a great philosopher', it is fair to assume that he is echoing *Science and the Modern World* (PR, 11). A new 'stage' has been constructed by a 'shock' that now prohibits making reference to prior thoughts. It is, of course, never impossible to refer to prior thoughts, and Whitehead himself never stops speaking about seventeenth- and eighteenth-century categories. Such references, however, are always on a different level, with transformed stakes. James's essay and the *Discourse on Method* produce precisely the same thing: a profound irreversibility.

Pragmatism is not a theory. Nor is it a vision of the world. It is simply a method. James repeats this many times.

> At the same time [pragmatism] does not stand for any special results. It is a method only. But the general triumph of that method would mean an enormous change in what I called in my last lecture the 'temperament' of philosophy.[3]

The aim of the second lesson of *Pragmatism* is to lay out the pragmatic method as both a technique and an instrument.[4] This method has essentially two aspects:[5] it is, first of all, a method of *evaluation*. It offers a technique that allows ideas and philosophical systems to be evaluated and put to the test. As such, '[t]he pragmatic method is primarily a method of settling metaphysical disputes that otherwise might be interminable'.[6] Such a testing of ideas does not pass through any criteria immanent to those ideas, it does not analyse what they are, their essence or their definition, but rather their *effects*. The fundamental rule of the pragmatic method *qua* technique of evaluating ideas is formulated for the first time by C. S. Peirce in his famous article 'How to Make Our Ideas Clear'.

> Consider what effects, that might conceivably have practical bearings, we conceive the object of our conception to have. Then, our conception of these effects is the whole of our conception of the object.[7]

The importance of this rule, in such a straightforward formulation, is that it eliminates all the problems linked to the truth and adequation of ideas. Truth becomes an event that 'happens' to ideas,[8] predicates them. An idea is not said to be true or false; it *acquires* truth, it *becomes* true. The process by which truth is ascribed to ideas is precisely what Peirce terms 'effects'. As a result, to evaluate an idea is to *follow a movement* in experience, to discern a 'difference in practice', since what is decisive is '[n]ot where it comes from but what it leads to',[9] the movements it settles down in, the trajectories it produces. The method is presented as *an art of effects*.

Pragmatism, as shown above, is a method of inventing ideas. This interest in an idea as it is constructed, or fabricated, though never given, derives from a proposition concerning the nature of experience. This proposition is stated by James, though variations can be found in Peirce or Dewey: 'What really *exists* is not things made but things in the making.'[10] If experience is not constituted by 'ready-made' things, any adequation of ideas to experience emerges from a virtual state, a 'doing' understood as a movement or direction. If an idea is to be linked effectively to what exists in an intermediary state of becoming, then the idea itself has to be produced in such a way, it must express its own power of becoming. The correspondence between an idea and the real is shifted. What was originally a correspondence is transformed into an analogy between two movements: *the movement of things in the process of becoming and the movement of ideas as they are constructed*. The important thing is to remove the question of method from a set of relations, of rules aimed at a knowledge organised entirely around the discovery and evidence of a supposedly direct correspondence between experience and ideas.

These two aspects of the pragmatic method determine the point of irreversibility that James introduces to the question of method. And it is from this point, the immediate presupposition of *Process and Reality*, that speculative thought will be constructed as a method. Whitehead, however, never minimises the differences between pragmatism and speculative thought. Recognising his debts to the pragmatists, to whom he adds Bergson, he hints, nevertheless, towards areas of genuine disagreement with pragmatism, in the form of a reserved statement regarding certain 'misunderstandings' in pragmatism's reception.

> I am also greatly indebted to Bergson, William James, and John Dewey. One of my preoccupations has been to rescue their type of thought from the charge of anti-intellectualism, which rightly or wrongly has been associated with it.[11]

The Constraints of the Speculative Method

Speculative thought is situated in the wake of the irreversibility produced by the pragmatists on the question of method. It distinguishes itself from pragmatism, however, by insisting on the necessity of situating the question in a form of radical rationalism.

> Speculative Philosophy is the endeavour to frame a coherent, logical, necessary system of general ideas in terms of which every element of our experience can be interpreted. (PR, 3)

This definition has been the object of numerous commentaries, and there seems to be unanimous agreement that every book on Whitehead should start by clarifying the meaning of these components and determining their range. This approach, however – taken by most studies of *Process and Reality*[12] – catches the reader in the trap of thinking that what is at stake is the construction and account of a *system of ideas*, a complex group of abstract propositions linked together in an organic system. *Process and Reality* would then be an attempt to construct a coherent, necessary and logical system of thought able to account for experience. This makes sense as a vision, of course, but in my view the work's originality lies elsewhere. It is to be found, instead, in the line that connects this proposition to the definition of speculative thought that precedes it. Rather than an account of Whitehead's system, then, this proposition is a continuation of the speculative method. It describes not general categories but *the method's constraints*. There is a big difference between the two.

The passage can be read in a manner continuous with the speculative method, and its effects can be evaluated. Things are reversed: what Whitehead terms a 'system of ideas', or what he more frequently refers to as a 'scheme of ideas' (so as to maintain its dynamic and open character), is not a first element, something we begin with, but rather the method's aim, its goal or object. Consequently, when Whitehead writes that this system of ideas must be 'necessary, logical, coherent', and that it must allow every element of experience to be interpreted, these are not the characteristics or qualities of a system, but rather form a technical account of the method's constraints.

These constraints form two distinct groups, though they mutually and concretely implicate each other to the extent that it is difficult to separate them without changing their individual natures. The first group is composed of the terms 'necessity', 'logical' and 'coherence'. These are what Whitehead calls the 'rational' constraints, and he speaks of the rational dimension of the scheme only when emphasising one or more of these three constraints. The second group is expressed by an extremely general proposition: 'every element of our experience'. Later, but still in the same section,

Whitehead clarifies the constraints within which this extensive account of experience should be undertaken: 'The empirical side is expressed by the terms "applicable" and "adequate"' (PR, 3). Just as the first three terms define the 'rational' dimension, the two others, by contrast, determine the 'empirical' dimension of the scheme. Between them, these five concepts comprise the entirety of the speculative method's constraints. These are the only concepts we have. No proposition beyond these five requirements can be added to the method to impose a more general meaning on to it.

Clearly, a recovery of these terms, loaded with a history that overdetermines their meaning, cannot be undertaken without some risk. Adequation seems to refer directly, as its etymology attests, to the idea of a correspondence in the sense of a 'resemblance' or an 'image'. As for application, it evokes the idea of an 'implementation', of a confrontation with a 'real' to which ideas would supposedly have to be connected. It is clear that speculative thought sits very badly with the classical terms of adequation and applicability that presuppose an exteriority in the relation between reality and ideas. This is why Whitehead clearly clarifies the highly distinctive meaning of adequation: it means, rather, that 'the texture of observed experience, as illustrating the philosophic scheme, is such that all related experience must exhibit the same texture' (PR, 4). This definition of adequation is fundamental in so far as it expresses an imperative: *the same texture has to be exhibited.*

The scheme has to be 'adequate', not to observed experience, which would return us to the idea of a correspondence to a state of affairs or an event, but instead to *related experiences*. The relation is transversal: adequation becomes *a relation between areas of experience.* It involves the construction of relations, connections, links that allow experience to be thought as a multiplicity of connected elements. If this relation were simply given in the real, adequation would serve no purpose – it would be enough simply to point to the space containing such a link. It must instead be invented, constructed. The range of the constraint embodied by this transformed idea of adequation could seem excessive: it implies that 'there are no items [of experience] incapable of such interpretation [by the scheme]' (PR, 3). Because of this primary constraint the scheme works through *generalisation*. It advances step by step, crossing areas of convergence, establishing new relations, attempting to realise a structure from which no element would be excluded a priori.[13]

But the scheme would fall into an empty generality of meaning if the empirical dimension were conceived only in terms of a descriptive generalisation produced by adequation. In order to be complete, the scheme's capacity to account for experience has to be permanently put to the test. This is the principal role of the constraint of applicability, presented instead as a constraint of validity: it is a question of making sure that 'some elements of experience are thus interpretable [by the scheme]' (PR, 3). The distinction between the two – adequation and application – is not clearly visible at first sight. Indeed, if *all* elements of experience have to be interpretable by the scheme's construction (adequation), then obviously *certain* elements will be (application). In other words, the difference is not a logical one, since in purely logical terms it goes without saying that the '*all*' includes the '*certain*' as a particular case. The difference, however, is pragmatic. When we speak of adequation we focus on the scheme's *relational capacity*, whereas when we speak of application we emphasise its *capacity to account for* the particularities of experience.

An extravagant faith in the speculative method might be detected here. Isn't adequation an overly ambitious constraint with respect to the use of concepts? After all, with the speculative method every experience has to be interpretable as an 'illustration' of a scheme of ideas. Does this not demonstrate an extreme faith in the capacity of philosophy to account for all of experience? And when Whitehead writes that '[t]here is no first principle which is in itself unknowable, not to be captured by a flash of insight' (PR, 4), can we not ask what permits the possibility of such adequation? No philosophical system can claim to answer the scheme's empirical demands, but there is doubtless *an optimism of entitlement* in *Process and Reality*: *experience must be conceptually thinkable.* There should, for every problem, exist an adequate manner in which it can be posed, a manner that would be applicable to all related facts.

And yet '[p]hilosophers can never hope finally to formulate these metaphysical first principles. Weakness of insight and deficiencies of language stand in the way inexorably' (PR, 4). Limits of intuition are relative, they are linked to a finitude of thought that stops every totalising apprehension of experience in its tracks, no matter how intuitive. The limits of language go much deeper, and are linked intrinsically to the activity of philosophy itself: '[w]ords and phrases must be stretched towards a generality foreign to their ordinary usage' (PR, 4). Philosophy's tool is language pushed to its limits. Philosophy, then, inherits all of language's limitations. The limitations of grammatical forms (subject–predicate), fashioned on the basis of particular habits, are adapted with difficulty to the practice of abstract generalisation. Whitehead sees abusive generalisations of linguistic forms in most metaphysical problems.[14] This is why one of the first tasks of philosophy should be to pay particular attention to the legacy it receives from everyday language and to explicitly construct a language that would belong to it, an abstract syntax.

> Every science must devise its own instruments. The tool required for philosophy is language. Thus philosophy redesigns language in the same way that, in a physical science, pre-existing appliances are redesigned. (PR, 11)

Rationalist Trust

How could adequation think experience as a 'particular case' in the scheme of ideas if the scheme of ideas were not in some way already present? How would application allow the elements of experience to be detected if a tendency towards this were not already underway? The empirical constraints presuppose – though without any chronological priority – the other dimension of the scheme, its 'rational' dimension. Whitehead only indicates the differences between the two dimensions through their respective constraints. If, however, we disregard their continued schematic interconnectedness in favour of the qualities belonging to each, a split appears concerning the status of ideas themselves. If the empirical dimension exemplifies an awareness of the relations between ideas and experience, then the rational dimension can be characterised as an economy of ideas in their pure abstraction. No longer used to generalise experience, thought rather accounts for the specificities of ideas as such.

The constraint of necessity demands that the scheme contain 'in itself its own warrant of universality' (PR, 4). The concept of necessity has shifted: it no longer concerns a foundational element of the scheme, a first principle or category from which the others would be derived. It emerges instead from the *relations* between ideas. Ideas refer to nothing but their own arrangement within a system constituted by their reciprocal implications. Only in this way can a scheme be 'necessary'. Again, however, this constraint is purely technical. It describes nothing that we usually understand by 'necessity', nor by 'how ideas must be coordinated'. It simply states that a successful construction is one that carries within itself, immanently, guarantees of its own universality. And it is supplemented by another constraint, the ambition of which is far more modest and entirely dependent on the others: logic. This constraint lacks its own content. It takes no part in the construction of the scheme. Its function is clearly delineated: to make sure the scheme contains no 'inconsistencies'.[15] Again, logical inconsistency, the most famous expression being the principle of non-contradiction, concerns only the relations between ideas. It does not demand we step outside of ideas to appraise their consistency, simply that we contrast them with each other. Its main aim, then, is to avoid inconsistency. And yet, for Whitehead, inconsistencies have minimal importance: 'logical contradictions, except as temporary slips of the mind – plentiful, though temporary – are the most gratuitous of errors; and usually . . . are trivial' (PR, 6). What matters lies elsewhere: '[t]he verification of a rationalistic scheme is to be sought in its general success' (PR, 8).

The constraint that truly defines the rational dimension, however, is that of *coherence*. Coherence has a completely different function from logical consistency and should not be mixed up with it. Coherence produces harmonies, it links ideas together, forming an organic system of interconnected elements. Whitehead introduces this constraint in an analysis in which he attempts to give it its full significance: '[t]here is not a sentence which adequately states in own meaning. There is always a background of presupposition which defies analysis by reason of its infinitude.'[16] Any proposition that matters, whatever its form and however purified it is, includes presuppositions. Such presuppositions can be diverse – definitions of words, intentions of the speaker, context and so on – and are mobilised within the proposition with greater or lesser degrees of subtlety. And each one of these presuppositions leads to others, to the extent that, step by step, presupposition by presupposition, the movement leading back from the proposition to what it requires is infinite. A universe 'harmonises'[17] within each proposition, giving it meaning, and 'no entity can be conceived in complete abstraction from the system of the universe' (PR, 3).

Whitehead's insistence on philosophy's presuppositions can be explained by his refusal of a particular vision of thought and philosophy. He is opposed to what he terms 'incoherence', defined as 'the arbitrary disconnection of first principles' (PR, 6). Incoherence crystallises around a definition of thought as encompassing sets of distinct and autonomous propositions that could each be treated in their own simplicity and self-evidence, allowing their truth to be evaluated. One might think of Descartes' *Rules for the Direction of the Mind*, of which Whitehead writes that its 'system makes a virtue of its incoherence' (PR, 6) since it generalises, at the level of the organisation of thought itself, its definition of substance as that which 'requires nothing but itself in order to exist'.[18] This is the principle posed at the heart of all propositions, and it is this principle that Whitehead struggles with when he makes coherence – the *expression* within each proposition that reaches to infinity – the speculative scheme's chief constraint. This explains why he opposes the Cartesian principle with *a relational vision* in which the existence of each element requires the universe as a whole: 'Whenever we attempt to express the matter of immediate experience, we find that its understanding leads us beyond itself, to its contemporaries, to its past, to its future, and to the universals in terms of which its definiteness is exhibited' (PR, 14). It must be possible to construct a 'systematic universe to supply [. . .] requisite status' to every proposition, every idea and every concept (PR, 11).

The Function of Method

The scheme's two dimensions never stop linking themselves together, recomposing themselves according to the elements they take into account: coherence varies according to the generalising experimentation of adequation, testing it and redeploying it in other modes; application is directed by 'rational' constraints that give it a 'manner' of being in relation to experience, 'a way of dealing with data'.[19] Necessity is transformed by coherence, forcing it to travel along other routes, as well as by empirical constraints that demand its continued reformulation. Whitehead describes this perpetual movement of method in more intuitive language with the help of a metaphor.

> The true method of discovery is like the flight of an aeroplane. It starts from the ground of particular observation; it makes a flight in the thin air of imaginative generalization; and it again lands for renewed observation rendered acute by rational interpretation. (PR, 5)

The speculative method is dynamic, unable to stabilise itself once and for all. This is a result not only of the empirical restrictions already mentioned (intuition and language) but also since the method's internal relations require permanent reconstruction. The dynamism belongs to the ideas themselves. It is this set of connected constraints, empirical and rational in equal measure, as if ideas were constructed simultaneously in two dimensions, that Whitehead calls *speculative thought*.

This set of operations and constraints, processes that both produce knowledge and require that any knowledge produced be connected back to those constraints, is interesting from the point of view of speculative construction. What, however, is its function? In the passage cited above the question was posed at the level of experience, of a certain relation to experience in a mode of descriptive generalisation. If the scheme has an importance and a function, then it emerges through the relations it holds with experience. Ultimately, '[t]he elucidation of immediate experience is the sole justification for any thought; and the starting-point for thought is the analytic observation of components of this experience' (PR, 4). To elucidate,[20] to investigate experience. Sometimes the verb 'to disclose' is used. *Speculative construction has as its function, then, the elucidation of immediate experience.* But Whitehead also uses a more technical term at the heart of speculative thought: interpretation.

It might be said that the concept of interpretation sits uncomfortably with the radical rationalism announced above, and one finds such statements throughout *Process and Reality*. It is important to remember, however, that Whitehead is working with a redefined notion of interpretation: 'By this notion of "interpretation" I mean that everything of which we are conscious, as enjoyed, perceived, willed, or thought, shall have the character of a particular instance of the general scheme' (PR, 3).

Far from returning to a kind of relativism that would imply the absence of any criteria of relevance and truth, interpretation finds itself within *strict constraints* that have a surprisingly rational aspiration: *the scheme must be able to interpret everything*. Within this new structure, interpretation is the element that brings the empirical and the rational into continual communication, everything that is felt as 'enjoyed, perceived, willed, or thought' is connected to what is constructed. In this sense interpretation is not one constraint among others. The scheme is what makes the interpretation of immediate experience possible. That is, it is what allows us to shift what is experienced in a particular case to a relational universe in which all elements of experience can interact, traversed by infinite lines of reciprocal presuppositions. Speculative thought is a method, a method of interpretation.

Notes

1. Translator's note: see Whitehead, *Procès et réalité*, p. 45.
2. Whitehead, *Science and the Modern World*, p. 143.
3. James, *Pragmatism*, p. 31.
4. 'Theories thus become instruments, not answers to enigmas, in which we can rest' (James, *Pragmatism*, p. 32).
5. My description of the pragmatic method here owes much to David Lapoujade, *William James*.
6. James, *Pragmatism*, p. 28.
7. Peirce, 'How to Make Our Ideas Clear', p. 258.
8. Worms introduces a fundamental contrast between James and Bergson according to their different relations to truth. He writes: 'for James truth is superior to reality, whereas for Bergson reality is superior to truth, but . . . for both there is a fundamental gap between the two, between truth and reality' (Worms, 'James et Bergson: Lectures croisées', p. 54).
9. James, *The Will to Believe*, p. 24. Cited in Lapoujade, *William James*, pp. 59–60.
10. James, *A Pluralistic Universe*, p. 117.
11. Whitehead, *Process and Reality*, p. xii.
12. The lack of awareness of the importance of *Process and Reality*'s first sentence is clear in a number of works devoted to Whitehead. This implies that the idea of starting a reading by describing the scheme is far from given. Rather than challenging works as important as those by Ivor Leclerc (*Whitehead's Metaphysics*), William Christian (*An Interpretation of Whitehead's Metaphysics*) or Wahl (*Vers le concret*), all of which have marked, each in its own way, the legacy of *Process and Reality*, I would simply note that these works all lack a definition of speculative thought as a *method*. Notable exceptions to this approach are Félix Cesselin (*La philosophie organique de Whitehead*), who dedicated his first chapter to the question of method and, more recently, Isabelle Stengers (*Thinking with Whitehead*), whose work has established crucial analogies between speculative method and mathematics within the framework of constructivist thought.

13. In *Modes of Thought*, Whitehead writes: 'Philosophy can exclude nothing' (p. 2). This is a new return to James's radical empiricism: 'To be radical, an empiricism must neither admit into its constructions any element that is not directly experienced, nor exclude from them any element that is directly experienced' (James, *Essays in Radical Empiricism*, p. 22).
14. Whitehead sees in the 'subject–predicate' grammatical form the foundation of one of the principal metaphysical orientations. According to him, this structure determined the ontological principles of Greek thought: 'Greek philosophy had recourse to the common forms of language to suggest its generalizations. It found the typical statement, "That stone is grey"; and it evolved the generalization that the actual world can be conceived as a collection of primary substances qualified by universal qualities' (PR, 158). Modern philosophy inherits these principles in a new language, a new terminology, but remains unable to break with the subject–predicate difference: 'All modern philosophy hinges round the difficulty of describing the world in terms of subject and predicate, substance and quality, particular and universal' (PR, 49). Metaphysics can also be described as what Whitehead tends to oppose himself to, that is, a certain manner of writing and speaking, a certain language.
15. Whitehead returns to this question of 'inconsistency' in *Modes of Thought*. According to him: '[t]he foundation of logic upon the notion of inconsistency was first discovered by Professor Henry Sheffer . . . Professor Sheffer also emphasized the notion of pattern, as fundamental to logic. In this way, one of the great advances in mathematical logic was accomplished' (Whitehead, *Modes of Thought*, p. 52). He sees, here, a return to the Spinozist definition of finitude, a definition according to which the finite 'is that which excludes other things comparable to itself' (*Modes of Thought*, p. 52).
16. Whitehead, *Essays in Science and Philosophy*, p. 73.
17. One might think of Leibniz's expression: '[i]t can even be said . . . that all things harmonize . . . and that eyes as piercing as God's could read in the lowliest substance the universe's whole sequence of events' (Leibniz, *New Essays*, p. 55).
18. Cited in PR, 6.
19. Whitehead, *Adventures of Ideas*, p. 223.
20. In *Adventures of Ideas*, Whitehead links this practice of elucidation to the construction of harmonies between different parts of experience. Philosophy's function is 'to coordinate the current expressions

of human experience, in common speech, in social institutions, in actions, in the principles of the various special sciences, elucidating harmony and exposing discrepancies' (Whitehead, *Adventures of Ideas*, p. 222).

3

Creativity as Ultimate Principle

What is an Ultimate Principle?

Whitehead writes that '[i]n all philosophic theory there is an ultimate' (PR, 4). It would be difficult to find a case study in *Process and Reality*, a description of the functioning of a philosophy's ultimate principle. The only example that can be found is a succession of terms: 'In monistic philosophies, Spinoza's or absolute idealism, this ultimate is God, who is also equivalently termed "The Absolute"' (PR, 7). Whitehead says no more of it. But how does this ultimate work? Why has he been able to identify it with the 'absolute' or with 'God'? These questions remain unanswered. It is possible, however, to reconstruct a discourse of the absolute by taking a detour to connect a few separate elements.

Bergson poses a similar question in *The Creative Mind*. Departing from Berkeley's proposition '*esse est percipi*', he attempts to identify the function and highly distinct status that it occupies in *A Treatise Concerning the Principles of Human Knowledge*. It is possible, according to Bergson, to connect the four principles of Berkeley's system to pre-existing propositions by retracing his heritage across the history of philosophy. There would be, then, an idealist thesis defined by the proposition that 'matter is a cluster of ideas', a nominalist thesis according to which abstract ideas are reduced to words, a 'spiritualist' or 'voluntarist' thesis characterised by the will and that affirms the 'reality of minds', and finally a 'theist' thesis. Bergson writes: '[n]ow, nothing would be easier than to find these four theses, formulated in practically the same terms, among the contemporaries or predecessors of Berkeley'.[1] The first is close to Malebranche's occasionalism, the second to Hobbes and Ockham, the third to Duns Scotus and, on certain points, to Descartes, and the last to certain theologians.

Although this is interesting and necessary work, however, it tells us nothing about the particular emphasis Berkeley gives each thesis, an emphasis that turns *esse est percipi* into a key proposition that lets it radically break with the philosophers from whom it inherits its features. There is something in the proposition *esse est percipi* that distinguishes it from all of the philosophers listed above and appears to give it a particular status, a centre, what Bergson calls *an 'impulse' point*, as if its impulse connected it to every part of the system. Everything happens as if each problem were directed by and adapted from a new perspective, set in motion in such a way that it allows us to identify his idealism with nominalism and his nominalism with voluntarism. This identification, or fusion, of the elements of his philosophy underscores the importance of the proposition *esse est percipi. The simplicity of an intuition expresses itself in Berkeley's entire philosophy*, for Bergson, and it is this intuition that defines the singularity of his thought.[2]

The example of Berkeley corresponds almost exactly to Whitehead's understanding of the 'ultimate'. Whereas Bergson, however, sees an intuition to be recovered, a more or less implicit presupposition that could be brought to light through the work of elucidation, Whitehead sees a proposition or principle to be *posed*. 'Ultimate' means 'that which is the furthest beyond', the 'most remote', or simply the 'last'. The ultimate in a succession would be either the first term, the term before which the series as such does not exist, or the last, the term from which one either exits the series or transfers to another one. In this reasoning, the ultimate is the limit – the first or final term – of an explanation or demonstration. When Whitehead speaks of an ultimate he is referring to a technical concept in argumentation: in every demonstration there is a first element that is not the object of that demonstration but which must first be *posed* if a succession of reasons is to follow.[3] One must, in short, *begin by accepting the arbitrariness of all beginning*, that is, by accepting that something must be posed that has meaning only as the 'impulse' of a series.

Since the 'ultimate' is not the object of a definition as such but serves rather a *function* it cannot be considered in itself. An ultimate allows something to be constructed of which, each time, it is the first term and the 'impulse', just as Berkeley's *esse est percipi* is located at the beginning of a multiplicity of successions, a series of signs, of images, of sensations, of terms of knowledge and so on. But this does not imply that, *qua* 'ultimate', it cannot be the object of evaluation. It means, rather, that such evaluation cannot define truth criteria; it cannot return to the adequation of states of things nor to demonstrative consistency. All evaluation, rather, is essentially pragmatic: how does such an ultimate hold a group of concepts and propositions together? What are its effects in experience? What sorts of successions does it initiate? What impulses does it give to thought? I am in full agreement with Stengers when she makes the ultimate in the Whiteheadian sense an essentially pragmatic issue connected to the *technical* construction of problems:

> The ultimate, in Whitehead's sense, cannot, as we recall, by any means be identified with any form of transcendence, in the sense that any kind of sublime or intrinsically unthinkable character would be attached to it. If there is transcendence, it is a 'technical' transcendence. The ultimate is not the judge of problems and opinions, but is relative to the way the problem is framed and therefore liable to change along with the problem.[4]

Process and Reality's Ultimate Principle

This general approach to the 'ultimate' can now be explained with regard to *Process and Reality*. The question, clearly, cannot be substituted or overstepped. The issue, rather, is to know what the ultimate is within the framework of *Process and Reality*'s own method, to determine the ultimate element organising its construction. Whitehead gives this the name 'creativity'.[5] '"Creativity" is another rendering of the Aristotelian "matter", and of the modern "neutral stuff". But it is divested of the notion of passive receptivity, either of "form", or of external relations' (PR, 31).

The correspondence between concepts like 'matter' or, here, 'neutral stuff', makes no reference to any conceptual proximity, to any underlying identity, but instead to a function: 'creativity' has the same function, it occupies the same position as Aristotelian 'matter'. A major opposition, however, arises as soon as the meaning of 'creativity' is clarified: 'it is divested of the notion of passive receptivity'. What the two concepts share is the ambition to account for existence as such, in its most fundamental tendencies. Everything, in so far as it exists, from the smallest particle of matter up to the most stable forms of existence, expresses 'creativity' in some way. 'Creativity' becomes the great principle of existence just as, in Aristotelian thought, the 'first substance' is the principle of the existence of all things. It follows that *Process and Reality*'s structure must be evaluated from this primary tendency: *the production of novelty*.

It is as if every concept, every problem, were an 'actualisation' or an 'incarnation' of creativity. To think 'actualisation' is not to think the relation between a principle and its effects, between the foundation and the founded, as if the elements 'expressing' the ultimate could simply be deduced from it, as if the group of concepts which implement the ultimate were already contained, virtually, within it. As an ultimate principle creativity has no other meaning. It has no content: it is *a tendency of thought and nothing more*. It is not what is gained by 'transcending' concepts, a transcendence that would be aimed at a superior 'adequation', a pure intuition. It is simply the manner in which concepts express each other (PR, 7).

If creativity, however, like every ultimate, cannot be approached in itself, if it can be considered only locally, it nevertheless has a general meaning. As a *posed* principle, it guides all definitions: '"Creativity" is the universal of universals characterizing ultimate matter of fact' (PR, 21). Here, then, another way of expressing what has been described above as the essential characteristic of every ultimate principle becomes clear, namely as a tendency of concepts, a principle of all philosophical principles. That is, 'the many, which are the universe disjunctively, become the one actual occasion, which is the universe conjunctively' (PR, 21). Through creativity, then, the passage from the many to a conjunction that Whitehead calls an 'actual occasion' – which is, as I will show, the fundamental element of the speculative structure – can be understood. For the moment, however, we do not know what the many, or what conjunction, is. Every element of creativity will have to be constructed and explained, but at this point its general tendency can be clarified: it is *an advance towards conjunction*.[6] It is up to the actual entity to articulate the concepts that will turn it into a 'creature of creativity'. Three consequences follow from this.

1) Creativity gives precedence to novelty. To qualify creativity, Whitehead writes that '[t]he universe is thus a creative advance into novelty. The alternative to this doctrine is a static morphological universe' (PR, 222). Novelty is not a quality, a secondary dimension added to the things that constitute our experience, in the way that a pre-existing thing could be said to be new if and

when it is characterised by a new quality. When he makes novelty a fundamental trait of creativity and creativity the ultimate principle of his philosophy, Whitehead places the production of novelty on a completely different plane. Creativity concerns everything. The same expression of novelty is found in the most apparently simple elements and the most composed, in the most ephemeral and the most durable, in a particle and in the most complex of societies. Whitehead expresses this with a principle: 'the [. . .] world [. . .] is never the same twice' (PR, 31). That the universe is a production of novelty in each of its aspects does not require demonstration – this is what is *posed*. It is necessary, however, to construct the universe's philosophical, which is to say its technical and conceptual, expression.

In placing the production of novelty at the centre of his 'cosmology' Whitehead situates himself in a tradition of which the thought of Schelling, to which he refers in *The Concept of Nature*,[7] is undoubtedly one of the most important moments. In his 'Introduction to the Outline of a System of the Philosophy of Nature' Schelling develops what he calls a 'speculative physics'. The chief idea of this 'speculative physics' is to take objects of experience in their becoming, as 'products' that have first been 'produced'.[8] Every object, in so far as it exists, is an object produced by 'the original productivity of Nature'[9] from which it is derived and that it expresses. In identifying nature with productivity, an identification that marks the origin of his physics, Schelling explains how novelty and change can appear as first and ultimate realities.

> The philosophy of nature does not have to explain the productive power of Nature; for if it does not posit this as originally in Nature it will never bring it into Nature. It has to explain the permanent. But the fact *that* anything should become permanent in Nature, can itself only be explained by that contest of Nature *against all permanence*.[10]

This passage has much in common with the idea of creativity as ultimate. What Schelling calls 'productivity' is posed 'primitively' as an immanent activity. The task of philosophy is not to explain this 'productivity', a productivity that escapes all definition or deduction since it is the condition of everything and is required by all explanation, but rather to explain the 'permanent'. It could, therefore, be said of Whitehead what Émile Bréhier wrote of Schelling:

> Philosophy studies being in its becoming, in its formation, not as it is given. This becoming is not a pseudo-becoming in which new beings would appear as the logical conclusion or the new combination of pre-existing beings. It is a real becoming, the birth and development of something new.[11]

Schelling, however, tends to identify becoming with nature itself *qua* general movement. As a result, he is led to reduce the importance of *that which is produced* and to search – in an intuitive manner – for the conditions of an adequation to nature in its self-motion. His speculative physics, which should rather have been concerned both with what is given in experience in its irreducibility and the necessity of accounting for what is given by artificially *posing* a principle that would permit making more general expressions, is reduced to an attempt to comprehend nature as it is in itself, to construct a more accurate adequation. It can be said, then, that what Whitehead retains from Schelling is the demand of 'speculative physics' as a 'manner' of relating to experience – taking into account each element in its irreducibility and the necessity of structuring a scheme of artificial ideas with the goal of accounting for them – but that what he rejects is the valorisation of nature as a principle, *the naturalisation of the ultimate in all forms.*[12]

2) Creativity does not derive from any other being. 'Creativity', then, can be distinguished from all being, in the sense that it is, as Leclerc writes, 'the basic activity of self-creation generic to all individual actual entities'.[13] I will return to actual entities in the next chapter, but what should be highlighted for now is the status of this ultimate that, according to Leclerc, is for Whitehead the 'basic activity of self-creation'. The first consequence of this particular status of creativity, and what it shares with every ultimate principle, is that it cannot be derived: it is not creativity *of* something, of God or a primary substance, but nor is it creativity *for* something, with a view to such and such a being. If it is generic, it is required by all things at the same time, as it requires nothing but itself to exist. It is in this sense that the notion of 'self-creativity', the *causa sui* of creativity, should be understood. It is without foundation, since 'whatever its avatars, [it] will always communicate with what is "without reason"',[14] and at the same time it is *indifferent* or *neutral* with respect to its 'actualisations', that is, to groups of actual entities, to what exists.

To understand the at once underived and neutral character of creativity it can be contrasted with a concept that it resembles: creation. What distinguishes 'creation' from 'creativity'? First, 'creation' presupposes a moment of non-existence, a difference, more or less clear cut, between what already exists and what does not yet exist. Creation implies a gesture, an intervention or a group of principles producing an existence. The concept of creativity, with which things are constructed out of pre-existing elements, refuses the radical path from the creator to the created as well as the impression that, strictly speaking, nothing exists before the act of creation.[15] Everything comes from something, and whereas 'creation' sees cuts and ruptures, creativity sees an operativity that might be described in terms of resumptions, transformations of prior modes of existence, new relations, passages from latent to actual forms. With creativity there is never a true or absolute beginning but always a process in the making, new arrangements made from others, new lines of emergence spread over old ones.

Besides, creativity is distinguished from creation on the question of a creator. 'Creator', here, should be taken in the widest sense: it could be God, but it could also be any original motive power, anything that could be said to be *creative* of another thing. The concept of 'creation' generally rests on a distinction or a difference between that which is created and the creator itself, placing the latter in a relation of transcendence and exteriority in relation to what it creates, going so far as to ideologically valorise such a difference. The 'creator' must be clearly differentiated from the object of its creation to maintain the impact of the act of creation itself, its irreducible character. By maintaining the importance of the act as an absolute beginning, wholly dependent upon that which performs it, the concept coheres with a vision of the genesis of things as a passage from inexistence to existence. 'Creativity', on the other hand, is a relation that implies no radical difference between the object created and the action of the creator. Even God in *Process and Reality* is defined as a creature of creativity and is understood on the basis of phrases such as 'Immanent Creativity, or Self-Creativity' which avoid 'the implication of a transcendent Creator'.[16]

3) Creativity exists only under conditions: it is 'described as conditioned' (PR, 31). Although one and neutral, it exists only 'in virtue of its accidents' (PR, 7), that is, its actualisations. Since it is one, it is transcendent, but since it consists of a *multiplicity* of

operations inherent to all that exists, it is immanent. These 'actualizations' are 'actual entities' or 'actual occasions', each resembling 'the creativity of the Universe' in 'its own completeness, abstracted from the real objective content which is the source of its own derivation'.[17] In this sense, any approach that would attempt to think creativity as an object in itself, to turn it into a substance or even just a simple movement responding to its own logic, has to be rejected. Creativity exists only in its actualisations and every analysis must necessarily depart from such actualisations. A form of realism or universalism is here connected to the most radical forms of nominalism: *only the singular exists.*

A series of propositions can therefore be established from the above to construct the speculative scheme of *Process and Reality*: 1) there is something ultimate in every philosophy; 2) in Whitehead's philosophy this ultimate is creativity; 3) creativity is the production of convergence out of disjunction; 4) it differs from its actualisations but exists only by way of them.

Notes

1. Bergson, *The Creative Mind*, p. 93.
2. 'Finally the whole is brought together into a single point, which we feel could be ever more closely approached even though there is not hope of reaching it completely. In this point is something simple, infinitely simple, so extraordinarily simple that the philosopher has never succeeded in saying it. And that is why he went on talking all his life . . . All the complexity of his doctrine, which would go on ad infinitum, is therefore only the incommensurability between his simple intuition and the means at his disposal for expressing it' (Bergson, *The Creative Mind*, pp. 88–9).
3. For an analysis of the senses of the term 'ultimate' in *Process and Reality*, see the interesting article by Maurice Élie, 'Sur l'*Ultime*. A propos d'une catégorie de *Procès et Réalité*'.
4. Stengers, *Thinking with Whitehead*, p. 254.
5. Saint-Sernin, in an essential work dedicated to Whitehead's thought, fails to avoid a certain ambiguity when dealing with the question of the relation between the 'ultimate' and 'creativity'. Taking up the same passages that I have quoted above, Saint-Sernin writes: '[i]n all philosophical theories there is an ultimate element which is never graspable directly and which can be discovered only through its "accidents"' (Saint-Sernin, *Whitehead*, p. 35; translator's note:

my own translation). Despite fully agreeing with this passage, which summarises the stakes of the ultimate very well, it seems to me too ambiguous to conclude that 'this ultimate element, according to Whitehead, is creativity'. The risk here lies in interpreting the history of 'ultimates' in philosophy as a history that expresses or is aimed at creativity. On the contrary, we should take Whitehead literally: in *every* philosophy there is an ultimate that belongs to it and that cannot be confused, even in the form of an analogy, with creativity in the sense given to it by *Process and Reality*. All retrospective readings of creativity that would see it at work in previous philosophical systems will necessarily lead to misunderstandings, not only regarding the specificity of process metaphysics but also concerning the form of speculative thought.

6. 'The ultimate metaphysical principle is the advance from disjunction to conjunction, creating a novel entity other than the entities given in disjunction' (PR, 21).
7. See Whitehead, *The Concept of Nature*, p. 57.
8. See Schelling, *First Outline*, p. 202.
9. Schelling, *First Outline*, p. 202.
10. Schelling, *First Outline*, p. 206.
11. Bréhier, *Schelling*, p. 14. (Translator's note: my own translation.)
12. References to Schelling are common in pragmatism, most notably in the works of Peirce and Dewey. In 'The Doctrine of Necessity Examined', Peirce similarly places the question of the production of novelty in nature at the centre of his metaphysics. He writes: '[b]y thus admitting pure spontaneity or life as a character of the universe, acting always and everywhere though restrained within narrow bounds by law, producing infinitesimal departures from law continually, and great ones with infinite infrequency, I account for all the variety and diversity of the universe, in the only sense in which the really *sui generis* and new can be said to be accounted for' (Peirce, 'The Doctrine of Necessity Examined', p. 41).
13. Leclerc, *Whitehead's Metaphysics*, p. 84.
14. Stengers, *Thinking with Whitehead*, p. 254.
15. For an account of the differences between 'creation' and 'creativity', see François Jullien, *Procès ou creation*, specifically the chapter 'ni créateur ni création'.
16. Whitehead, *Adventures of Ideas*, p. 236.
17. Whitehead, *Adventures of Ideas*, p. 212.

4

Actualising Creativity

The Relation between Existence and Creativity

The description of the general components of creativity led to the following key proposition of *Process and Reality*: creativity exists only through its actualisations. But what does it mean to exist through actualisations? First of all, staying true to the literal meaning of what is posed, it implies that, strictly speaking, *creativity does not exist*, or at least, it does not exist outside of the operation of actualisation. As a result, creativity cannot be treated in itself, it cannot be considered in its own being, since this would presuppose its existence. This leads to a highly distinctive approach to existence with regard to creativity: it appears that existence is something added to creativity, something that happens to it within a process. If it were internal, Whitehead would have said that only *one* of the forms of creativity's existence is to be found in its actualisations, which would imply other forms belonging to it, relativising existence through actualisation. Whitehead's proposition, however, is the opposite: creativity's existence is related to its actualisations; it is drawn towards distinct things. It could be said, then, and without getting too involved in this point for now, that there is a difference between creativity and existence. It is a common metaphysical error, according to Whitehead, to confuse the ultimate with existence, making the latter into an attribute of the ultimate (in the form of substance, for instance, or of an atom).

Whitehead, however, often speaks of 'its' actualisations. This leads to a genuine paradox: existence is external to creativity and yet it is located only in elements internal to it (*its* actualisations). This paradox emerges from the manner in which the logical order of speculative construction gives an impression of following the order of existence. If it is necessary to *pose* creativity as the original moment in speculative thought, it is important to recall that it is posed only with a view to something else, namely, existence itself. If the order of exposition is followed too closely there is a risk of making existence into a problem posed by creativity. *Process and Reality*, however, is not a book about creativity. It is an attempt to speculatively construct a scheme of ideas, at the centre of which is existence. Creativity is therefore necessary to account for this existence, but it cannot be the element from which every other question would be posed. In short, Whitehead's proposition should be inverted to escape the paradox: *actualisations*, or existences, are primary. The ultimate must allow existence to be interpreted, rather than offering a derivation or foundation of it.

But there is another problem, here, one that is central to *Process and Reality*: what is meant by existence as *actualisation*? Actualisation comprises part of a definition of existence. If the ultimate must be posed, it is not to follow an extended chain of consequences that would derive from it – a particular approach to existence – but rather to set up the question of *actualisation*. And it is from this point of view that Whitehead invents a concept to link 'existence' with 'actualisation': the concept of an actual entity. 'The positive doctrine of these lectures is concerned with the becoming, the being, and the relatedness of "actual entities"',[1] he writes in the preface of *Process and Reality*. This justifies the approach I have chosen to take. The concept of the actual entity aims to give an account of existence itself, not of *modalities* of existence, existence as it is perceived or imagined. Whitehead calls for a metaphysics of existence that cannot be reduced to an exploration of perceptual consciousness or lived experience. It is a question of understanding what is meant when it is said that an individual, a thing or even an idea *exists*.

> 'Actual entities' – also termed 'actual occasions' – are the final real things of which the world is made up. There is no going behind actual entities to find anything more real. (PR, 18)

Occasionally Whitehead compares this notion to classical ideas of 'substance', 'monads' or even 'atoms'. These comparisons do not imply that these are the same thing in different terms – the actual entity is radically different from 'substance' in its classical meaning, or from the monad – but rather that what they share is an ambition to construct a concept of *existence in itself*. It could be said that what unites them is their function or ambition rather than their definitions or characteristics. Leclerc should be recognised as the first to establish a link between the conception of existence in *Process and Reality* and the search for a 'total fact' deriving from the Aristotelian approach, the 'general metaphysical category of "that which is"'.[2] For Leclerc, the 'that which is' is at the centre of the foundational question of a general metaphysics inherited by Whitehead, to the extent that every dimension of this metaphysics has to be linked to and integrated with the concept of an 'entity'.

For this reason, the concept of the 'actual entity' allows a kind of ontological limit or starting point of existence to be attained. It is impossible to ascend towards a reality from which it could be derived. The actual entity is an absolute beginning in the order of existence, beyond which 'there is nothing, merely nonentity – "The rest is silence"' (PR, 43).

What is Meant by 'Actual Entity'?

Most studies of *Process and Reality*, if they recognise the central place accorded to the notion of actual entities, tend not to linger on the terms themselves, as if the fact that Whitehead used two terms, each loaded with a singular history, was of little importance. This gives the impression that other terms could have been used, and that the central question should be located not in the manner in which the concept of the actual entity is constructed but in its function or definition. In a way, this choice is justified: Whitehead himself fails to clarify the terms 'entity' and 'actuality'. And yet, by reflecting on the meaning of these two terms, and the implications of the fact of their relation, by occupying the perspective from which they have been developed, the meaning of the concept 'actual entity' can be clarified.

The term 'entity' was in common use at the time *Process and Reality* was written. It is a familiar term in modern logic, with which Whitehead was well acquainted. It names, schematically, every thing, individual, class or group that can be treated as an individual reality, an abstraction effected by its concrete determinations and spatio-temporal inscription. It is a pure abstraction, a totality emptied of all content. In this sense, it is analogous with the concept of '*entitas*' found in scholastic thought.[3] And yet, some years before *Process and Reality*, the American philosopher

Lloyd Morgan, who had a link to Whitehead, took up the concept of entity in his main work, *Emergent Evolution*.

Morgan's significance comes from his translation of the concept of entity from logical to metaphysical language. In his work entities become the first element of a metaphysics of experience that resists all forms of abstraction. What Morgan retains is the sense of an individual existence that, although it can be broken down into parts or elements, forms an indivisible totality. Entities, in his metaphysics, correspond to everything that can be expressed by demonstratives such as 'this' or 'that'. Whatever the domain under consideration, whether physical, biological or social, there exist concrete and individual first elements that can only be expressed with a 'that' and are irreducible to their qualities or characteristics, a kind of pure 'this'.[4] It is the entity's resistance to abstraction that Whitehead inherits in *Process and Reality*.

Whitehead takes from Lloyd Morgan, first, the idea that an entity is a totality of undivided existence that avoids the distinction of whole and parts. It is an individual substance. Second, Whitehead inherits Morgan's radical monism in which nothing exists except entities. Whitehead makes this monism into one of the principal demands of speculative thought: '[t]he presumption that there is only one genus of actual entities constitutes an ideal of cosmological theory to which the philosophy of organism endeavours to conform' (PR, 110). The concept of entity becomes generic: all realities, all domains of existence, will need to respond to the monist demand according to which nothing exists apart from actual entities.

> They differ among themselves: God is an actual entity, and so is the most trivial puff of existence in far-off empty space. But, though there are gradations of importance, and diversities of function, yet in the principles which actuality exemplifies all are on the same level. (PR, 18)

Having clarified 'entity', it can now be asked what is meant by 'actual'. What does Whitehead mean when he explains that only 'actual' entities exist? What bearing does the concept of actuality have on the concept of the entity as so defined? Again, if its meaning is to be clarified then the history that gives it meaning, the history from which it retains its key elements, cannot be ignored. 'Actual' is the translation of the Latin *actus*, meaning 'action', 'act' or 'activity'. *Actus* is what 'acts', in contrast to what is 'passive'. Inertia, then, the act of resisting an action, remains an 'activity' in the classical sense of the term. To say that a thing is 'actual' means that, in this first sense, it is 'active', it 'exerts' in some way. It is for this reason that one of the translations of *actus* is 'efficiency', that is, the influence of one thing on another, the transformation effected by a thing on something else. The causal relation is therefore one of its modalities, and what is called an 'efficient cause' one of its expressions. A cause is 'efficient' in so far as a thing performs an activity.

A second meaning, however, is added to the first sense of activity or efficiency that does not appear directly in the term *actus*. We should return to its origin: *actus* is the Latin translation of Aristotle's *energeia*. *Energeia* is the activity or efficiency of a thing as it *actively exists*. It is opposed, therefore, to 'possibility' but also to 'capacity' and 'potentiality'. The active being of *energeia*, inherited by the notion of 'act', is opposed to potentiality, to *dunamis*. In the *Metaphysics* Aristotle gives a famous example: 'we say that potentially, for instance, a statue of Hermes is in the block of wood and the half-line is in the whole, because it might be separated out'.[5] Potentiality is expressed by the eventuality, the possibility, the 'could be', whereas the act is defined as that which exists fully and *actually*. Not a hypothesis, a possibility of being, then, but a real existence. The 'could be' of *dunamis* is opposed to the 'this is' of *energeia*. Here, then, is the full meaning of the actual: it is a real activity, not a hypothetical one, and it concerns an efficient existence, not a possibility of existence.

If an entity is defined as the 'whole essence' of an individual, what it is in itself, then the introduction of the notion of actuality implies that the whole essence is situated in a *real activity* that constitutes that individual. As Leclerc writes:

> It is not 'existence as such, in the abstract; it is the existence of a particular, a 'that'. Moreover, the 'that' which is in question is the that which is possessed of 'full existence', the that which exists 'in and of itself'.[6]

Displacing Actualism

The notion of entity is placed into an essential relation with the notion of actuality, a relation that determines its meaning.

Whitehead, then, situates his project in a tradition that could be called actualist.[7] This actualism emerges from an Aristotelian proposition: 'apart from things that are actual, there is nothing – nothing either in fact or in efficacy' (PR, 40). With this proposition Aristotle does not mean to diminish the significance of other dimensions of existence, including being in potential (*dunamis*), but rather to refer these other dimensions of existence to being in actuality (*energeia*). In book nine of the *Metaphysics*, on the subject of potentiality, Aristotle writes that 'actuality is prior to potentiality'.[8]

This anteriority is of several orders: first, the actual is prior to the concept. As such, by '"capable of building" [it is meant] that which can build, and by "capable of seeing" that which can see, and by "visible" that which can be seen'.[9] The 'capacity', the 'possibility' of seeing presupposes an 'exercise' of vision. According to the 'concept', there is an anteriority of 'exercise' over 'disposition'. In this way 'the formula and the knowledge of the one must precede the knowledge of the other'.[10] Second, actuality is prior with respect to time. Every being is an actualisation of a potentiality: the man exists potentially in the child, the wheat in the seed, and so on. But what exists potentially is 'posterior in time to other actually existing things, from which they were produced. For from the potential the actual is always produced by an actual thing, e.g. man by man, musician by musician.'[11] An actuality is preceded by another actuality in which it exists potentially, but this potential derives from a prior actuality, to the extent that, chronologically, the actual is prior to potential. Third, the actual is prior according to substance. When it is said that a thing is possible or potential it is because we have in mind its being in actuality, without which, according to Aristotle, we would not be able to say what is possible. We can say that the man exists potentially in the child because we have an idea of the man's being in actuality, his form. The actual is therefore prior to substance; the actual is presupposed as a finality.

When Whitehead describes the act as that which determines and gives meaning to the entity, he emphasises a certain pre-eminence of the actual over the potential. This does not mean that he joins with Aristotle regarding the precise *forms* of the actual's anteriority, or concerning the specific examples from which Aristotle constructs such anteriority, but simply that he sees in the actual the first element of existence. The pre-eminence of the actual does not imply, however, a rejection of potentiality or of the possible. Nor, incidentally, does it imply as much for Aristotle. In *Process and Reality*, possibility – real and general potentiality – occupies a central place in what Whitehead calls the categories of 'explanation'. But actualism refuses a treatment of the possible *qua* possible, the construction of a domain of potentiality that could be treated and considered outside of the actualities that constitute existence. How could it be possible to conceive of potentiality if it didn't *come from* somewhere, from an actual existence, if it weren't engaged in a local process, that is, if it weren't *situated in* existence? It is this attention to the *attached*, *engaged* or *situated* character of all possibility that characterises actualism and that interests Whitehead: possibilities are always found, not in the background, but in actual relation to what exists at a determinate moment.

Whitehead, however, distances himself from classical actualist thought on one essential point, namely, the implicit identification of actuality with simplicity, in the sense of an *individual* relation, in the precise sense of the term, an 'indivisible' reality. Turning this characteristic into the primary element of existence forces one to consider the actual as unformed, as that which contains no other reality within itself. To be in actuality means, then, to be exactly what one is, no more and no less. Such a valorisation of the actual can be found in empiricism. Locke, for instance, writes that 'when the Mind considers *Cajus*, as such a positive Being, it takes nothing into that *Idea*, but what really exists in *Cajus*'.[12] The idea of *Cajus* is an idea in actuality, since it contains nothing other than *Cajus* in his current form, that is, in his simplicity. For Whitehead, this search for simplicity is important in the framework of philosophical construction. On the other hand, the confusion that can emerge between simplicity and the elements that constitute experience can be 'disastrous'. Already in *The Concept of Nature* Whitehead had elaborated the risk inherent in this search for simplicity:

> We are apt to fall into the error of thinking that the facts are simple because simplicity is the goal of our quest. The guiding motto in the life of every natural philosopher should be, Seek simplicity and distrust it.[13]

The entire task from then on is to discover how to preserve the essence of the actualist approach, how to emphasise the *individuality* of the ultimate elements of existence, while at the same time refusing to reduce this individuality to simplicity.

The initial definition of 'actual entity' is a whole and individual essence (entity) determined by an 'activity', an 'act' or 'action' (actual). This individuality does not necessarily need to be posed in the configuration or form of a thing. It can simply be the type of action performed by that thing. Far from demanding that we leave actuality behind to understand individuality, the notion of the actual entity brings individuality back into actuality itself. An 'actual entity', an existence, is what it does. And every action is, as I will show, essentially a relation: something acts on something else and the relation is the action itself. Actualism tends to link the being's 'individuality' to its 'quality' and, consequently, it tends to favour 'the Aristotelian dominance of the category of "quality" over that of "relatedness"'.[14] In *Process and Reality* relation takes precedence over quality, the latter appearing only as an emergence, an effect, in a space of relation. And actual entities, in the sense in which they are *actualisations* of creativity, are essentially relational beings, *unifications* from which individuality emerges according to the manner in which they act. Individuality therefore passes from quality to efficiency, from simplicity to relation. An entity is individual in this way; it has 'significance for itself . . . [it] functions in respect to its own determination' (PR, 25).

The Actualist Recovery of the Principle of Reason

Process and Reality's general definition of existence can now be clarified, a definition that nevertheless remains to be properly constructed: *all existence is an actualisation of creativity and this actualisation is the becoming-actual of an entity*. This definition has one important repercussion for speculative thought. If, strictly speaking, only actual entities exist, then it follows that every question as to the 'reason' of a thing, its 'cause', would have to focus on one or more actual entities. This is a genuine principle that completes the speculative method, and Whitehead calls it the 'ontological principle': '[t]his ontological principle means that actual entities are the only reasons; so that to search for a reason is to search for one or more actual entities' (PR, 24).

All things find their 'reasons' in an actual entity, but actual entities have no reasons other than themselves. This principle appears to be the necessary consequence of the place given to actual entities, and by constructing it Whitehead touches on a fundamental idea in empiricism, the most explicit expression of which can be found in Hume's *An Enquiry Concerning Human Understanding*, which is why Whitehead also calls it 'Hume's principle'. What is this principle? It can be found on the final page of the second section of the *Enquiry*, the section on the 'origin of ideas':

> When we entertain, therefore, any suspicion, that a philosophical term is employed without any meaning or idea (as is but too frequent), we need but enquire, *from what impression is that supposed idea derived*?[15]

Hume presents this principle as a 'method'. When we encounter a philosophical term we have to locate the impression from which it is derived. This is a critical method, since its function is to bring concepts and ideas back to their foundations in experience. What Hume is aiming at, here, is the radical calling into question of the 'principle of sufficient reason', the principle according to which everything that exists has a foundation, a cause that explains it. Existence is what must be explained, it is that which poses a problem. And yet Hume inverts the question. It is not impressions that must be explained or founded – they are given and impose themselves through their self-evidence – but rather it is ideas, principles, concepts and essences that one must be able to explain as deriving from impressions. Whitehead, however, distances himself from Hume's method. First, the ontological principle in *Process and Reality* is not brought back to a theory of ideas and impressions, to an exploration of the understanding and its functions. Consequently, the ontological principle is not a critical principle that aims to offer a foundation for things. It simply indicates that every question relative to any existing thing refers the questioner to one or several given actual entities that can be deduced from nothing but themselves.

What Whitehead takes up in his own method – the speculative method – is the distribution between what explains and what must be explained: it is from an existence in actuality, efficient and individual, that the reasons for other dimensions of experience must be explained. This means, particularly, that 'although other types of entities do exist, they are (*i.e.* exist as) either "ingredients in" actual entities, or . . . "derivative from actual entities"'.[16] The great Humean split between 'impressions' and 'ideas' is transformed into a difference between entities as actual – the only ones that exist – and other entities, individual things not in actuality, such as potentialities or virtualities. Every entity has a *situated existence*, whether as an 'ingredient' in entities in actuality, that is, as elements comprising an activity, or as 'derived' from them. As such, '[e]verything must be somewhere; and here "somewhere" means "some actual entity"' (PR, 46). The meaning of the ontological principle can therefore be clarified: 'there is nothing which floats into the world from nowhere. Everything in the actual world is referable to some actual entity' (PR, 244). In short: everything has a relation with what Whitehead describes as a creature of creativity, and this situated character of existence is termed the 'concrete'.

Notes

1. Whitehead, *Process and Reality*, p. xiii.
2. Leclerc, *Whitehead's Metaphysics*, p. 54.
3. See De Freiberg, 'De ente et essentia', and, more specifically, the first section entitled 'De la signification de l'étant et de l'entité'. De Freiberg takes up the contrast between the *ens* and the *entitas*. The *ens* would contain 'everything in itself, in its extension, according to the thing and according to the meaning' (De Freiberg, 'De ente et essentia', p. 165; translator's note: my own translation). The *entitas* implies 'the same modes of consideration which have been defined for the subject of the *ens*', the only difference being 'that what is said of the *ens* in a concrete manner must be said of the *entitas* in an abstract manner, that is to say, the entity should be taken in the mode of abstract meaning' ('De ente et essentia', p. 169).
4. According to Leclerc, 'the term "actual entity", in its primary sense, signifies the general metaphysical category of "*that* which is"' (Leclerc, *Whitehead's Metaphysics*, p. 53). Wahl makes of it one of the principal orientations of pragmatism and its anti-Hegelianism. Taking Hegel's critique of the abstract and general character of the 'this', he writes that pragmatist philosophy, 'by placing the accent on the *mine*, on the *here*, the *now*, on all those elements of designation that can be seized by thought only by denaturing them', forces us to see the 'inanity of the Hegelian critique . . . It demands the right to the immediate' (Wahl, *Vers le concret*, p. 3; translator's note: my own translation).
5. Aristotle, 'Metaphysics', p. 1655.
6. Leclerc, *Whitehead's Metaphysics*, p. 20.

7. See Wahl, *Du rôle de l'idée de l'instant dans la philosophie de Descartes.*
8. Aristotle, 'Metaphysics', p. 1657.
9. Aristotle, 'Metaphysics', p. 1657.
10. Aristotle, 'Metaphysics', p. 1657.
11. Aristotle, 'Metaphysics', pp. 1656–7.
12. Locke, *Essay*, p. 319.
13. Whitehead, *The Concept of Nature*, p. 163.
14. Whitehead, *Process and Reality*, p. viii.
15. Hume, *Enquiry*, p. 16.
16. Leclerc, *Whitehead's Metaphysics*, p. 25.

Part II

The Speculative Approach to Existence: Process and Individuation

5

What is a Process of Individuation?

The Place of the Problem of Individuation

At the end of the last section we arrived at the following central proposition of *Process and Reality*: *existence is an actualisation of creativity*. This proposition is simply an *intuition* that can be formulated more explicitly by connecting the method to the ultimate principle. Like all intuitions, however, it lacks the technical and systematic structure that would allow it to occupy a central place within speculative thought. As an intuition it has, as Bergson writes, 'furnished an impulse, this impulse a movement'.[1] It is that movement that now must be described, the process by which existence as actualisation gives rise to a series of new problems, trajectories deriving from this initial impulse. Although the expression of the intuition was important, it was not enough: now we must follow its effects.

Recall its status. It is not descriptive. It does not try to say what existence would be in itself or to express the conceptual form of an observed existence met with in experience. It is defined by its function, namely, to allow for an *interpretation*. It is too early, of course, for the terms of this interpretation to be established as self-evident, a self-evidence that would presuppose the speculative movement itself in its entirety. So far we have only the axes of the problem's formulation, the tendencies and impulses on the basis of which the scheme is set up. Interpretation requires the mediation of the scheme. The fact, however, that the proposition is asserted in this particular context suggests that it is possible to treat it in its purest abstraction, as an ideality.[2] Its relation to the real and to experience can be held in suspense, postponed, in favour of its own meaning, what it implies 'in itself'. Although it speaks only of existence, then, it does so internally, within the scheme of ideas and, as such, within the constraints of speculative knowledge in its rational dimension: necessity, coherence and logic. Its consequences for experience, its 'adequation' and 'applicability', will take place in a field the terms of which will be clarified.

In its general and immediate form this has two important consequences: first, to speak of *actualisation* is to bring to light the 'processual' character of existence, the fact that, unlike Leibniz's monads which emerge through 'fulgurations',[3] existence is not given at once but is the outcome of a becoming or an emergence. However we understand the word 'actualisation', it carries the sense of a process, a set of operations. An entity does not exist directly but, to the extent to which it becomes an 'act', actualises itself, emerges as a true actual entity. The technical term to designate existence has great importance, here, since *it makes the fact of existing completely inseparable from the operation by which the existence emerges.* Following this, if actualisation is a question relative to the 'coming into existence' of an actual entity, then Whitehead's proposition establishes *Process and Reality* as contributing to a philosophy of *individuation.* Whitehead, of course, does not use the term 'individuation', preferring other, more precise concepts like 'concrescence', 'actualisation' or even 'prehension'. And yet one could speculate that these concepts, their characteristics and relations, all converge towards a philosophy of the individuation of actual entities. This hypothesis seems to open new perspectives, both on the history of the problem of individuation (from Aristotle to Leibniz) and on more recent reworkings that have no direct or explicit link to Whitehead's thought.[4] This is the hypothesis that will guide us along the speculative construction as a whole. It can be said right away: the axis upon which speculative coherence is constituted can be found in the term 'individuation'. This, then, requires some explanation, specifically of what is understood by individuation.

Gilbert Simondon, in a different and later context, developed an original philosophy of individuation that opens up possible ways of interpreting the stakes of *Process and Reality*. If it seems relevant to construct a link between Whitehead and Simondon, it is not on the basis of an analogy between their technical approaches to existence but rather thanks to the manner in which they locate the importance of the processual dimensions of the real. Their convergence is thematic rather than analogical; they share an 'organized network of obsessions'.[5]

In his two principal works, *L'individu et sa genèse physico-biologique* and *L'individuation psychique et collective*, Simondon invents what he terms an 'axiomatic of the human sciences' that he also terms an 'energetics'. From the first pages of *L'individuation psychique et collective*, he presents a similar ambition for the philosophy of individuation.

> In the search for a principle of individuation, a reversal must be brought about. The operation of individuation by which the individual comes into existence must be regarded as primordial. The individual, in its characteristics, reflects the workings, the rate and modalities of this principle.[6]

This is the declaration of a programme: 'a reversal must be brought about'. At times Simondon's approach can be identified as a manifesto, a kind of charter for the development *of a philosophy yet to be constructed*. Its originality is radical: it amounts to opposing a history of thought that, since Aristotle,[7] has revolved almost exclusively around the individual, a tradition that was translated into conceptual form by philosophy. What it fundamentally calls into question is not a certain concept, a particular way of considering the real and experience or a specific philosophical movement, but rather something that spans most of the concepts, movements and styles that constitute the history of philosophy. What Simondon tries to call into question is a *presupposition of philosophy*, an 'image of thought'[8] as powerful as it is rarely explicit, constructed on the basis of the 'ontological privilege given to the constituted individual'.[9] Crisscrossing philosophy is an interest in the completed individual, and philosophy's conditions, modes of existence, principles and categories are the conceptual generalisations of that privilege. Implicitly, *philosophy inherited a 'pre-philosophical' vision of the world in which experience and reality appeared as a set of individual, distinct and autonomous things*.

This paradigm of the 'constituted individual' requires two general conditions: first, the set of processes and movements by which things come into being must be refused or limited. If what matters is the individual, then there is a risk that individuation is seen as a mere phase, a more or less incidental stage that becomes legitimate only at the very end, in its finality, that is, in the individual. Individuation would then take the form of the very reality it was intended to explain. The second condition is the reduction of the set of the individual's connections and exchanges with a much larger environment in which it is entangled. Once again the individual is defined as a kind of inherent solidity, as autonomy or closure. Once proposed, all the links connecting it to its environment can be understood only as external relations.

The full importance of Simondon's call for a 'reversal' should be located, then, in his opposition to this paradigm of 'individual' reality, the efficacy of which, Simondon discovers, is at work up to the present. The reality that matters, the aim or foundation of all research, is no longer the individual but rather the processes of individuation, the set of regimes and operations by which things come to be. If, indeed, we start not with the individual but with these 'regimes', then primary reality, that upon which all the problems of existence have to be laid out, necessarily takes a different form and has different qualities to that of the individual. In so far as this primary reality precedes, is prior to or different from the individual, Simondon calls it 'pre-individual reality'. We should take this concept literally: *that which precedes the individual.* Whitehead calls it 'nature' and in so doing places himself into the pre-Socratic tradition of thinking *physis*.

> The meaning that the pre-Socratic philosophers gave to the word nature can be recovered: philosophers found in nature the origin of all varieties of being anterior to individuation. Nature is the reality of the possible, the *apeiron* from which Anaximander excluded all individuated form.[10]

Nature is this '*apeiron*', this map of the possible out of which the individual emerges but that it does not limit. In Simondon, it might be said, one finds a kind of naturalisation of individualisation, an insertion of individuation into a genuine philosophy of nature outside of which the individual being could not be understood. Individual being is what emerges from the possible without putting an end to it; what results from individuation is not the individual but the 'individual environment', a mixture, a hybrid of pre-individual nature and individuality. This is why the individual is never totally adequate to its individuality, why it is always 'more than one'.[11] It extends itself beyond itself into the nature connecting it to everything that exists. But it is also 'less than one', precisely because the self-identity of the individual can never be attained: the potentialities and possibilities spanning it cannot be channelled into a stable and homogeneous identity.[12]

The importance that Simondon assigns to regimes of individuation is close to Whitehead's thinking on the emergence of actual entities. The differences, however, are important. The reversal Simondon calls for leads towards a radical rejection of the individual in favour of a philosophy of nature in which the individual appears as a passing phase. In the passage cited above, Simondon uses an expression that reveals his wish to distance himself from a philosophy of the individual. He writes that the individual 'reflects' regimes of individuation, turning the individual into a mere 'reflection' or 'appearance'. Such a reduction of individual nature, if historically understandable, agrees very little with Whitehead's philosophy of 'actual entities' that are precisely *individual substances*. What's more, a return to a philosophy of *physis* is exactly what *Process and Reality* attempts to resist. There is, of course, an approach in Whitehead's philosophy that could indeed be connected to a philosophy of nature, especially in *The Concept of Nature*,[13] *Adventures of Ideas* and *Modes of Thought*. But this is true only in a highly particular sense and on the basis of different problems of which *Process and Reality* is neither the outcome nor the simple generalisation.

In *Process and Reality* the question of nature is no longer an original, primordial one, in the sense of allowing actual entities, existence or reality to be understood. Whitehead, unlike Simondon, has no desire to 'go beyond' the individual towards nature. In *Process and Reality*, nature is not *what explains*, it is not the source of the possible, but *what must be explained*: it is not 'pre-individual' but *fabricated*, *constructed* out of a multiplicity of individual beings. One might say that Whitehead attempts both to take up a philosophy of individuation very close in intention to Simondon's refusal of the classical notion of the individual while at the same time recasting the concept of the individual. His ambition is to construct a genuine philosophy of individuation that would no longer be based around a reality chronologically or ontologically prior to the individual – such as nature, or a kingdom of forms – to which the individual could be reduced.

The concept that has the function of establishing an *irreducible* link between the individual and individuation, between the actual entity and its actualisation, is concrescence. It is one of the major concepts of *Process and Reality*, just as important as actual entities or creativity. It is formed from the Latin *concrescere*, referring to the idea of a 'formation through the joining of several parts'. We speak of concrescence when physical or biological elements, each with their own existence, 'grow', 'develop' or 'amass' together. Other possible translations of the Latin are to 'stick together', to 'condense', or even to 'congeal'. Each of these expressions refers to the notion of a 'hardening', a 'development' of several elements or parts, each with their own individual existence. In its most basic form, concerning only two elements, two individuals are in the process of concrescence when their relation effects a transformation that gives rise to a common unity of existence. Whitehead also speaks of a 'togetherness': 'the "production of novel togetherness" is the ultimate notion embodied in the term "concrescence"' (PR, 21). Individuation, then, becomes a *congealing, a hardening of a link, a condensing* of a diversity of individual and heterogeneous elements.

Disjunctive Diversity: Potential Being

If individuation is indeed the congealing, the concrescence of diversity, then the first question is to know what diversity is. Whitehead develops the notion in the chapter entitled 'The Categorial Scheme', introduced as an 'anticipatory sketch of the primary notions which constitute the philosophy of organism' (PR, 18). The question of the philosophy of organism can be left to one side for now,[14] in order to go straight to the first elements of the description. What should be understood by an 'anticipatory sketch of the *primary notions*'? What, in *Process and Reality*, is a primary notion? It is tempting to see, here, a return to the classical question of foundational principles and categories on the basis of which the entire system could be unfolded as the logical outcome of the development of first elements. This approach to primary notions, however, would contradict the definition of speculative thought as a productive method of knowledge that follows the requirement of creativity. The chapter, then, targets something else, something that defines the status of the concept of the 'many'.

The chapter's ambition is to construct words, to provide a *vocabulary, a lexicon, a whole language*. As Peirce writes, 'I particularly approve of inventing new words for new ideas'.[15] For each new idea there is a corresponding suitable vocabulary that permits its true expression. In the case of *Process and Reality*, the problem is less the introduction of new ideas than *the possibility of expressing individuation*, inventing a language, a syntax. One might quite reasonably ask why such a language would be necessary. The main reason is that the problem of individuation, if correctly laid out, would have to avoid a substantialist approach, an approach that would be imposed through a particular form of language and syntax. Metaphysics – and particularly Aristotle's – is less a conception of the real than a *way of speaking*, with its own words, its own grammar and phrases. Resisting a metaphysics of substance, then, involves taking a distance from one's language by *inventing* new terms or *reappropriating* old ones. With these 'primary notions' Whitehead attempts ('an anticipatory sketch') to invent a purely technical and functional language. The words[16] that compose this language do not refer to definitions but rather to functions that have as their task the expression of individuation.

The question of the 'many' has to be framed in this artificial and technical setting:

> The term 'many' conveys the notion of 'disjunctive diversity'; this notion is an essential element in the concept of 'being'. There are many 'beings' in disjunctive diversity. (PR, 21)

This definition, on first sight, might appear redundant. It appears, after all, only to add to what is already known: that the 'many' is 'disjunctive diversity'. The two terms seem equivalent – isn't the 'many' already by definition a form of disjunction? This definition does clarify something, however: it directs and expresses the radical meaning of the 'many' in *Process and Reality*. Indeed, Whitehead never speaks of 'being in disjunction', as if the many were a state in which elements could simply be, coexisting in diversity, in what Bergson calls a quantitative multiplicity.[17] It is a question, rather, of *a process* of disjunction, a plurality to the extent to which it is disjunctive. The whole difference between 'being in disjunction' and 'disjunctive being' turns on the *activity* that characterises the latter. It is *actively* disjunctive. The term 'many' should therefore be reclaimed, but it should be given a quality lacking in its more classical senses. By introducing this new quality, the term passes from referring to a simple coexistence among plural things related only – and at the most – by the fact that they share the same space-time, to designating more active forms of differentiation expressed by verbs like to differentiate, separate, come apart or split up.

This, though, raises a second question: what composes this 'many'? Up to now we have had to use the word 'parts' to express the meaning of concrescence in a more 'intuitive' manner. But the word 'parts' is inadequate: it always, more or less subtly, refers to a whole/parts relation that presupposes existence as a totality or a unity, either prior to or following diversity. The 'many', however, is radically distinguished from a logic of totality that would attempt to recover and limit it. If nothing but actual entities exist, if the rest is 'silence', there can be only one possible way of expressing the many: to commit to its mode of existence in so far as *the many is the expression of the relations between actual entities*. For now we will have to make do with the extremely general indications provided by the term's components; any advance towards the term's implications would require putting the term to work within the speculative construction. If Whitehead poses the question of 'preliminary notions' before setting up his philosophy of individuation, however, it is because there is a rhythm to speculative construction, a rhythm that requires that these 'requisites' be constructed before they are put to use. Without this rhythm the philosophy of individuation would never properly unfold. It would lack the very language needed for its construction.

The 'many' can be thought in two ways according to the perspective taken: either in relation to itself or to another. The first sense, considered up to now, is abstract, a kind of speculative 'fiction'. Existence never presents itself in this form, in a pure, disjointed state. The second way of thinking is more important and more concrete. Because it is a question of individuation, here, the many has meaning as and when it is engaged within a new existence, within a reality in the process of realisation. In such situations it is called 'potentiality'. The many is the potential on the basis of which individuation functions, it is a possible-being. What is fundamental, here, is that potential is not a quality of disjunctive diversity in itself – disjunctive diversity's only quality, as we have seen, is being an operation of disjunction between entities in action. It is, however, a potential *relative* to new existences. In *Process and Reality* this relative character of potential is the great invention of the term 'many'.

As such, the relations between the potential and the real, between the possible and action, have been completely recast. The many, now, is composed only of beings in action. Nothing exists as a possibility, everything is efficient and real. This pure disjunctive actuality only becomes a potential when involved in an individuation: *the act becomes power, the real becomes the possible*. We should recall that there is a tradition in the philosophy of individuation that likes to differentiate between registers of being, turning individuation into the passage from one reality to another. Simondon, for instance, sees individuation as the passage from nature (the reality of the possible) to the individual (the actualisation of nature). Whether this passage is considered as a gain or as a loss, whether between possibility and reality something is added or taken away, the idea nevertheless stands that there is a difference in the being itself of the elements under consideration. This is precisely the difference Whitehead immediately refuses: the possible and the real, power and action, are not ontological differences or oppositions.[18] They are distinguished only by relations, they are relative to the situation in which the entities find themselves. Potential is action engaged in something else, in another entity. Whitehead summarises this in a passage that, for the moment, cannot be fully unpacked but should be taken as an indication of problems to come:

> in the becoming of an actual entity, the *potential* unity of many entities in disjunctive diversity – actual and non-actual – acquires the *real* unity of the one actual entity; so that the actual entity is the real concrescence of many potentials. (PR, 22)[19]

The term 'many' can now be clarified as the first dimension of being: it is the disjunctive multiplicity of actual entities in so far as they constitute a potentiality for new individuations.

Individuation as Capture

The passage from the many to the unity of a new entity operates through what Whitehead terms 'prehensions'. The term comes from the Latin *prehendere*, meaning 'to take', 'to capture' or even 'to appropriate'. It is the activity through which one thing takes or seizes another. And yet defining prehension as an act of seizure or capture risks limiting the process to a certain kind of activity, in particular to the banal activity of a subject capturing an object. The risk, then, lies in approaching the act of capture as a relation between two individuals already constituted before the capture itself, a relation between two otherwise autonomous and distinct realities. It limits the fact of *transformation*; the genesis brought about by the relation. This is why the concept of prehension is larger and more constitutive than the notion of capture or possession. It is concerned with a prior level, namely, individuation. *Appropriation*, then, is a more accurate translation of prehension, since it brings to light the movement of integration within a new existence. To appropriate is, indeed, to take or to capture, but it is also to make another reality one's own. There is a complex phenomenon that takes place in appropriation: a thing is captured, but captured in such a way that it becomes an object of another. No longer are we looking at two pure and distinct realities in external relation. No: we are looking at a transformative activity that changes and modifies the elements through the relation itself, moving elements into a sort of interiority.

There is no need, however, to think the process of appropriation as a process of homogenisation in which the other takes the image of what prehends it or is founded on its basis. To appropriate is not to force a resemblance. This would be possible only if that which prehends were already fully constituted and so capable of dominating the other, forcing it into an identification. Framing the question at the level of individuation, then, prevents the reduction of prehension to a kind of domination, a reduction to the 'same'. It is the emerging entity – which does not yet fully exist – that prehends other entities. Where, then, does homogenisation come from?

The definition of 'prehension' as the appropriation and integration into the individuation of a new actual entity allows the term 'actual entity' to be clarified. We have said that the notion of the actual entity was formed, in particular, from the idea of action, referring to notions of activity and efficiency. The term 'prehension', then, provides a technical clarification: *prehension is the constitutive activity of actual entities*. Actual entities are nothing but centres of prehension; their entire being is identified with their capture of other actual entities. They express neither essence nor finality, but simply operations of capture.[20]

Deleuze dedicates important pages to the concept of prehension in *The Fold*. He sees prehension as a key term for articulating the possible relations between Whitehead and Leibniz, and as an initial condition for a philosophy of events. The originality of Deleuze's approach lies in the connection he makes between the concept of prehension and that of disjunctive diversity. In proposing this connection, he distinguishes himself from the majority of Whitehead's readers who, if they did describe prehension correctly as the integration of other actual entities, did not necessarily raise the implications of that integration for disjunctive diversity. The chapter in which Deleuze develops the question of prehension is entitled 'What Is an Event?', attesting to the relation Deleuze sees between 'prehension' and 'events'. This chapter ends the second part of the book which focuses on the question of 'inclusions' – one of the fundamental dimensions of prehension understood as appropriation and integration. This is how Deleuze, in a highly general proposition, sets up the problematic space in which prehension is realised and gains its necessity: 'events are produced in a chaos, in a chaotic multiplicity, but only under the condition that a sort of screen [*crible*] intervenes'.[21] What Deleuze calls chaos corresponds precisely to what Whitehead calls 'disjunctive diversity' (PR, 21): a multiplicity of divergent potentialities, of disjunctions between actual entities. What he calls an 'event' corresponds to a new actual entity (despite the quid pro quo to which I will return). Having established the space of the problem, having restated that the question of prehension has no *raison d'être* outside of disjunction diversity, Deleuze traces the new entity's movement, or passage, out of chaos.

> Chaos does not exist; it is an abstraction because it is inseparable from a screen that makes something – something rather than nothing – emerge from it. Chaos would be a pure *Many*, a purely disjunctive diversity, while the something is a *One*, not a pregiven unity, but instead the indefinite article that designates a certain singularity. How can the Many become the One? A great screen has to be placed in between them.[22]

First, taken by itself, as its own reality, disjunctive diversity is an abstraction. It has meaning and effectiveness only when involved in a new concrescence. Next, the new actual entity – which Deleuze expresses with the wholly appropriate term 'certain singularity' – emerges from the many. Finally, this passage presupposes a 'screen' [*crible*]. The question of the 'screen' is fundamental. It is precisely what allows the passage from plurality to unity, *from chaos to a certain singularity*. Without this screen, chaos would remain disjunctive diversity without the slightest possibility: for Whitehead, as for Leibniz, possibility is always *the possibility of order or unity*. For Deleuze, prehension is a central concept that allows this screen to operate in disjunctive diversity, that is, to sort, to evaluate, to distinguish what the new entity is genuinely able to capture.

How, though, does prehension work? Let's stay with Deleuze's reading of Whitehead.

> Everything prehends its antecedents and its concomitants and, by degrees, prehends a world. The eye is a prehension of light. Living beings prehend water, soil, carbon, and salts. At a given moment the pyramid prehends Napoleon's soldiers (forty centuries are contemplating us), and inversely.[23]

Deleuze's reading is coherent: the many becomes a certain singularity through successive prehensions that, step by step, form a world proper to each, *a cosmos for each out of the soil of an initial and shared chaos*. Incidentally, this is the fundamental point of disagreement between Whitehead and Leibniz. For Whitehead what is shared is not the world as expressed by all monads[24] but rather a chaos created by everything that already exists, a multiplicity of potentialities. The world, on the other hand, belongs to every new existence.[25]

Although this reading – entirely directed as it is towards the question of the production of order out of diversity – uncovers key perspectives on *Process and Reality*, it is nevertheless structured around an initial confusion, a confusion that has serious consequences. Deleuze reads *Process and Reality* as a philosophy of 'events'. The notion that *Process and Reality* turns around the question of events is an important idea, one I would not want to contest. And yet, for me, the question is rather to know *where to situate events?* What are 'events' in *Process and Reality*? It is on this point that Deleuze, along with a certain number of French readers, makes the same mistake as Jean Wahl.[26] Their responses are similar: 'actual entities' are the technical expression of *Process and Reality*'s philosophy of events. This is clear for Wahl, it requires no problematisation. Here is what he writes on the subject of 'prehension', which resonates nicely with Deleuze's examples:

> The word prehension signals a beyond, it points towards the vectorial and, as the phenomenologists would say, intentional character of the

> concrete event. The essence of a real entity consists of prehension [. . .] It feels something 'there' and transforms it into a 'here'.[27]

Actual entities are events and their essence is to prehend. Defining events as prehensions, then, makes all sorts of generalisations possible: the eye prehends light, the pyramids prehend soldiers, the living prehend water, and so on. Events appropriate not just their past (their antecedents) but everything contemporary with them (their concomitants). They integrate into themselves everything that surrounds them and, step by step, the world in its totality. This vision of prehension is interesting, and it makes what Whitehead attempts to construct with the concept of prehension more intuitive and 'visible'. What makes it powerful, however, is exactly what it suffers from, namely, its over-reliance on intuitive examples: rocks, pyramids, soldiers. If Whitehead provides virtually no examples of prehension – and if he does, it is with extensive and repeated reservations – it is for a reason.

Actual entities are not events. The question of events has to be located elsewhere, in what Whitehead calls 'societies', to which we will return. Wahl, and Deleuze in his wake, confuses actual entities with societies, two things that *Process and Reality* basically opposes to one another. The former are essentially 'abstract', corresponding to nothing in experience; their level is 'microscopic' (PR, 128). The latter are 'concrete events', events we experience in perception, in our sensations, our desires and so on. They comprise a 'macroscopic' plane. The philosophy of individuation always takes place on a 'microscopic' scale, a scale that can in no way be an object of our experience, either in the form of an empirical or an intentional object. Wahl's and Deleuze's examples retain their force, of course, and yet, instead of explaining or accounting for prehension, they can be taken only as an incitement to take what Whitehead calls an 'imaginative leap', a jump into pure abstraction. It is technically incorrect to say that a 'rock' or a 'life' are 'prehensions'. Such examples do, however, evoke what needs to be thought at the level of actual entities.

What is a prehension, from a speculative point of view? It is, as Wahl saw, a question of a 'vector': prehension effects the passage from a reality 'over there' to a reality 'right here'. It makes disjunctive diversity pass into a new actual entity in its 'real internal constitution'.[28] This, of course, is a deeply Leibnizian idea: just as monads express the universe from a particular viewpoint, actual entities, one might say, prehend all the others from their own perspective. Keeping the necessary abstraction, it could be said of actual entities what Leibniz wrote on the subject of monads:

> every substance is as it were an entire world and a mirror of God, or rather of the whole universe, expressing it in its own way, somewhat as the same town is variously represented according to the different positions of an observer.[29]

In the case of actual entities, however, there is no expression of any prior or pre-given world, a world that could be expressed by its expressions.[30] There is simply the prehension of other actual entities and the constitution of a particular world through that prehension. For Whitehead, prehension expresses 'the activity whereby an actual entity effects its own concretion of other things' (PR, 52). Prehension, then, is essentially *unifying* in so far as it makes the 'there' move into a 'here', displacing other entities in the constitution of a new entity.[31] It produces a unity of existence, a perspective on the totality of what exists. When he defines perception, Leibniz uses an expression that could be transposed directly on to the level of prehensions: it is a 'multiplicity in a unity'.[32] The new entity is the emergence of a unity that captures, appropriates and engulfs the multiplicity of other actual entities.

A Different Economy of the Subject and the Object: Objectification

It seems that the concept of the actual entity has been taken, here, in two distinct ways: it describes the existences that comprise disjunctive diversity at the same time as it designates the entity in the process of individuation. In more Bergsonian terms, the actual entity seems to apply at once to a ready-made reality and to a reality in the process of being made. This ambiguity, however, is *the direct consequence of Whitehead's monistic requirement*, a requirement that rests on the principle that there can only be one form of existence. Doesn't this difference introduce a paradox, then? Is there not a contradiction between alternative demands: on the one hand there is a monist affirmation that only actual entities exist, and on the other a dualism of the entities that compose diversity and the new entity? In reality, however, this objection rests on the denial of precisely what is fundamental to the philosophy of individuation, namely, the becoming of being. Particular to this philosophy is the inclusion, in all of its questions, of *phases, moments or stages of existence*. A thing does not exist completely at one time but *accedes* to existence in a movement from which it is inseparable. In fact, identifying the concept of the actual entity with one phase alone, without becoming, would bring about a contradiction. To avoid having to explain the moment of existence in which we find ourselves at every moment, Whitehead introduces a technical distinction. Relative to individuation, 'objects' are those entities that already exist – those that compose disjunctive diversity – and 'subjects' are new entities. But we should not be misled, here: this is a purely functional distinction. In a certain relation of individuation, some entities are called 'objects' and others are 'subjects'. In both cases, however, they are actual entities, with the same requirements and form of existence.

Classical separations of subject and object erroneously 'reify' the link, they substantialise the operation. Once completed, this reification leaves us with a series of false questions and badly framed problems: how can a subject be in relation with an object? What qualities can be attributed to the one and the other? All of these questions disappear as soon as the crux of the problem is treated as purely *functional*: what role does such and such an entity occupy in an individuation? What does this function require and involve? How does the function of the one encounter the function of the other?

These new questions are unfolded from a central problem. If it is true that the prehending entity, the subject, captures existing entities, objects, including them in its own constitution, then the problem is to know what it means for an object to be present in a subject.[33]

In the vocabulary of the philosophy of individuation, 'objectification' is the name for the process by which an entity becomes internal to another. To construct the concept of objectification, to make it into a 'first' concept like the 'many' and prehension, Whitehead returns to Descartes. Surprisingly, he places the term 'objectification' in the direct legacy of the *Meditations*, a gesture even more unexpected for the fact that the term underlines *the relativity of the subject and object*, their constant intermingling, the impossibility of ontologically distinguishing them. In other words, Whitehead challenges the classical distinction between subject and object, but inscribes this challenge within the Cartesian tradition itself. After all, wasn't Descartes precisely among the most responsible for the modern systematisation of the radical separation between subject and object? It is clear that, by reiterating the Cartesian project, Whitehead profoundly transforms it.

Skimming through *Process and Reality*'s references to the *Meditations* on the topic of concrescence, it becomes clear that he never refers to the *Meditations* directly but rather to the *Objections* and particularly to the first. Descartes, responding to Caterus's criticisms of him, mobilises a scholastic terminology. He takes up the difference between two kinds of existence: formal and objective. This classical difference, found especially in Duns Scotus (to whom he refers),[34] is retranslated and reappropriated by Descartes in the terms of a distinction between ideas.

> the idea of the sun is the sun itself existing in the intellect – not of course formally existing, as it does in the heavens, but objectively existing, i.e. in the way in which objects normally are in the intellect. Now this mode of being is of course much less perfect than that possessed by things which exist outside the intellect; but, as I did explain, it is not therefore simply nothing.[35]

This passage allows us to clarify the distinction: the sun's *formal* existence is its existence 'in the heavens', that is, its existence in itself, outside of the intellect. Its *objective* existence is the very same sun, and yet it is the sun as it exists in the intellect. This distinction might seem classical: it refers, finally, to a difference between a reality 'in itself' and a reality 'in another'. And yet what interests Whitehead is the way Descartes seems to accept a double existence of ideas. Everything happens as if the sun were *simultaneously* located in its own place and somewhere else, in the intellect.

There is little doubt that Whitehead's reading is, to say the least, guided by an entirely non-Cartesian problem and, as a result, involves a complete transformation of Descartes' thought. Indeed, Whitehead transforms an example that Descartes limits to the status of ideas into a problem of existence: in formal reality there is an existence in itself, an existence which, at the same time, has an objective reality in other things. To displace Descartes' example of the sun on to the level of actual entities requires satisfying several conditions. First of all, it has to be removed from a problem that exclusively concerns ideas. Secondly, Descartes tends to minimise the importance of objective reality: he writes, at the end of the passage cited above, that the manner of being of these ideas 'is of course much less perfect than that possessed by things which exist outside the intellect', that is, formally. Unlike ideas, actual entities exist both objectively and formally without any hierarchy. Finally, Descartes bases his distinction on an example that obstructs a genuine conception of how something can exist in 'something else'. How can we imagine the sun lying truly *inside* the intellect without this being a metaphor or a mere idea?

Having laid out these conditions, we can return to the distinction at the level of the theory of individuation. Descartes' 'formal' existence becomes the existence of those entities that compose disjunctive diversity, that is, the mode of existence of the objects of concrescence. 'Objective' existence, however, is the existence of those very same objects but now *within the subject*, as if every entity were multiplied by the prehensions performed on it.

This distinction radically calls into question what Whitehead terms the 'principle of simple location'.[36] This principle, which pervades the modern sciences in particular, rests on the notion that the real is made out of elementary particles localisable at precise space-time points, that such particles can be situated at such a point in space and at such a moment. 'Objectification', however – the proliferation of the objective existence of actual entities by way of prehensions – implies a multiplicity of localisations at the same moment. Individuation, then, can be explained as the continuous passage from formal to objective being.

The distinction between the 'objective' and 'formal' is not the only thing that Whitehead takes from Descartes. He completes this distinction – with the same limits, reservations and displacements that we raised on the subject of the formal and the objective – with another idea that he places at the foundation of a genuine 'theory', an idea that, once again, is found in the *Meditations* and concerns the position of what Descartes calls 'feeling':

> Let it be so; still it is at least quite certain that it seems to me that I see light, that I hear noise and that I feel heat. That cannot be false; properly speaking it is what is in me called feeling (*sentire*); and used in this precise sense that is no other thing than thinking.[37]

If Whitehead cites this passage from the *Meditations* it is because Descartes affirms the unquestionable nature of feeling: we can doubt the existence of things, but not the fact that we feel them. Descartes expresses this certainty in a seemingly paradoxical formula: it is certain that it seems. Certainty concerning feeling, doubt concerning the objects of feeling. Noise, light, heat, in so far as they are interpretations of feelings, cannot refer to anything existing, and yet the intellect cannot doubt the existence of such feelings within itself. It is this certainty of feeling that Descartes identifies with thought itself.

What is it about this passage that interests Whitehead? Definitely not the way Descartes relates it to his system as a whole. Whitehead is quite clear on this point: 'I find difficulty in reconciling this theory of ideas (with which I agree) with other parts of the Cartesian philosophy.'[38] What he does take away from Descartes' short passage, however, is a definition:

> A feeling is the appropriation of some elements in the universe to be components in the real internal constitution of its subject. The elements are the initial data; they are what the feeling feels. (PR, 231)

This definition is clearly very close to the definition of prehension as the capture of other entities. Does this mean they are synonyms? Does Whitehead not merely provide a new term for the reality of being as capture, an idea he has already developed elsewhere? A priori they are identical: they share the same qualities and functions. And yet the examples Whitehead uses and the manner in which he defines them reveals that they are, in fact, used in different contexts. Whitehead uses 'prehension' to underline the operations of appropriation and capture by which 'disjunctive diversity' attains consistency. It is indeed used to explain the individuation of a subject but from the perspective of the appropriated 'diversity'. On the other hand, when Whitehead wishes to account for this diversity as it is experienced, as it is integrated by the subject, he prefers the term 'feeling'. We have already encountered this perspective shift regarding disjunctive diversity. It is as if the perspective taken by each term – prehension and feeling – foregrounded different aspects which are in no way contradictory or opposed but rather relative to the way in which the problem is set up. If it is a question of knowing, from the entity's point of view, the relation of that entity to its 'ingredients', then the term 'feeling' is used. If it is a question of knowing how those entities are integrated, then the term 'prehension' is more appropriate.

The problem of 'feeling', then, refers finally to the question of the subject, to its genesis and mode of existence.

Notes

1. Bergson, *The Creative Mind*, p. 91.
2. Among the important theses of Isabelle Stengers's book on Whitehead is the new relation she establishes between the speculative manner by which Whitehead performs the construction of the scheme and his past as a mathematician: '[if] Whitehead had been and remained a mathematician, it is precisely in so far as, for him, thinking has nothing to do with goodwill. He does not start out thinking God exists, aiming to show that all rational human beings with goodwill should accept the fact. He thinks in the mode of obligation, in the sense of being obligated by the problem he has constructed' (Stengers, *Penser avec Whitehead*, p. 17; translator's note: my own translation, since this sentence is omitted from the English version). The mathematical operations that Whitehead recovers are, first, the idea that every exercise of thought can occur only within postulated constraints on the basis of which a certain problem can be constructed. Then, a trust in a possible solution, the idea that an adequate manner of posing the problems can be built. Finally, a relation of creation, not discovery. As such, 'the art of problems refers to the freedom of the mathematician, for whom the solution to be constructed passes through the active indetermination of what the problem's terms might "mean"' (Stengers, *Penser avec Whitehead*, p. 27; translator's note: my own translation, since this passage is omitted from the English version). On this subject, see Cesselin, *La philosophie organique de Whitehead*. From the first pages of his work, Cesselin writes that 'the philosophical method, for Whitehead, has analogies with the methods of mathematics' (Cesselin, *La philosophie organique*, p. 11; translator's note: my own translation). It seems, however, that Cesselin limits the question of mathematics to that of simple deduction while Stengers is interested in the 'gesture' of the problem's construction.
3. In the *Monadology* Leibniz writes: 'Thus it may be said that a Monad can only come into being or come to an end all at once' (*Monadology*, p. 219).
4. I am thinking, here, especially of Raymond Ruyer, who in his key works, especially *Éléments de psycho-biologie* in 1946, developed a singular philosophy of psycho-biological individuation on the basis

of notions of theme and survey [*survol*] that deserve to be related to Whitehead's thought. Ruyer does not ignore Whitehead's existence, making explicit reference to him in *La gnose de Princeton*, a work in which he analyses the gnostic 'metaphysics' of Princeton scientists. He sees in the critique of the 'simple localisation of matter', whose invention he correctly attributes to Whitehead, one of the fundamental sources of this movement. See Ruyer, *La gnose de Princeton*, pp. 100–1. Of course, one might also think of Gilbert Simondon, whose thought will be returned to at greater length in what follows.

5. Barthes, *Michelet*, p. 3.
6. Simondon, *L'individuation psychique et collective*, p. 12. (Translator's note: my own translation.)
7. The critique of Aristotle is recurrent in Simondon's work and takes the shape of a critique of hylomorphism in its most classical form: the taking form [*prise de forme*] of a supposedly passive matter. In hylomorphism Simondon sees the expression of an operation of thought also to be found in psychology (particularly in gestalt theory) as well as in sociology (in the relations between the collective and the individual). The relevance of his critique of Aristotle pertains, then, not so much to Aristotle's own thought but to the effects of the generalisation Simondon performs.
8. See Deleuze's chapter on the 'image of thought' in Deleuze, *Difference and Repetition*.
9. Simondon, *L'individuation psychique et collective*, p. 10. (Translator's note: my own translation.)
10. Simondon, *L'individuation psychique et collective*, p. 196. (Translator's note: my own translation.)
11. On this subject, see Combes, *Simondon: individu et collectivité*, particularly the chapter entitled 'Pensée de l'être et statut de l'un : de la relativité du réel à la réalité de la relation'.
12. 'In every domain the most stable state is that of death, a state of degradation in which transformation is no longer possible without the intervention of external energy' (Simondon, *L'individuation psychique et collective*, p. 49; translator's note: my own translation).
13. Whitehead is very explicit on this subject in *The Concept of Nature*. Nature is 'that which we observe in perception' (*The Concept of Nature*, p. 3). It is not a question, then, of proposing the 'reality' of nature in the work but rather of analysing our modes of experience of nature, that is, principally, our perception. Whitehead holds exclusively and explicitly to the perceptual experience of nature. He asserts this definition several times in *The Concept of Nature*. In

this way, he ends the chapter entitled 'Theories of the Bifurcation of Nature' by exposing this limit of the analysis of perceptual experience: 'It is difficult for a philosopher to realise that anyone really is confining his discussion within the limits that I have set before you. The boundary is set up just where he is beginning to get excited. But I submit to you that among the necessary prolegomena for philosophy and for natural science is a thorough understanding of the types of entities, and types of relations among those entities, which are disclosed to us in our perceptions of nature' (*The Concept of Nature*, p. 48). It would, therefore, be in vain to search for a metaphysics or a philosophy of nature in the classical sense of the term, that is, in the tradition of Spinoza or Schelling. Here, it is merely a question of accounting for 'perceived' entities. This remark on the object and limits of *The Concept of Nature* is extremely important for avoiding hasty comparisons to *Process and Reality*, which crosses the 'limit' laid out in the passage cited above, radically departing from perception and its models.

14. See Cesselin, *La philosophie organique de Whitehead*, and in particular the first chapter entitled 'La méthode philosophique'.
15. Peirce, 'On Signs and the Categories', p. 220.
16. Dumoncel relates Whitehead's demand for a new vocabulary to William James, who wrote to his brother Henry, 'I have to forge every sentence in the teeth of irreducible and stubborn facts' (cited in Dumoncel, *Les sept mots de Whitehead*, p. 11).
17. See Bergson, *Key Writings*, p. 69.
18. See the extremely interesting essay by Pierre Caye, 'Destruction de la métaphysique et accomplissement de l'homme (Heidegger et Nietzsche)'.
19. On the same page there is another definition of potentiality: 'the potentiality for being an element in a real concrescence of many entities into one actuality is the one general metaphysical character attaching to all entities, actual and non-actual'.
20. Whitehead, here, is in tacit agreement with another of Leibniz's heirs: Gabriel Tarde. In *L'Opposition universelle*, Tarde writes: 'The living being aims to appropriate the world, not to adapt itself to it' (*L'Opposition universelle*, p. 160; translator's note: my own translation). And he develops a similar theme in *Monadology and Sociology*: each monad 'draws the world to itself, and thus has a better grasp of itself. Of course, they are parts of each other, but they can belong to each other to a greater or lesser extent, and each aspires to the highest degree of possession; whence their gradual concentration;

and besides, they can belong to each other in a thousand different ways, and each aspires to learn new ways to appropriate its peers' (*Monadology and Sociology*, p. 57). Here we find an idea common to both Whitehead and Tarde: existence is defined as a constitutive activity of appropriation by which other elements are integrated. Both place questions linked to capture and appropriation, linked, in a word, to 'possession', at the centre of the real. They substitute a logic of appropriation for a logic of being. But the relations between Tarde and Whitehead are not limited to question of appropriation. Both thinkers have converging approaches (on issues of repetition, heritage, resumption and differentiations) to societies (physical, biological and social objects).

21. Deleuze, *The Fold*, p. 76.
22. Deleuze, *The Fold*, p. 76.
23. Deleuze, *The Fold*, p. 78.
24. See Deleuze, *Difference and Repetition*, p. 47.
25. 'For Whitehead (as for many modern philosophers), on the contrary, bifurcations, divergences, incompossibilities, and discord belong to the same motley world *that can no longer be included in expressive units*, but only made or undone according to prehensive units and variable configurations or changing captures. In a same chaotic world divergent series are endlessly tracing bifurcating paths. It is a "chaosmos" of the type found in Joyce, but also in Maurice Leblanc, Borges, or Gombrowicz' (Deleuze, *The Fold*, p. 81).
26. Jean Wahl's influence on Deleuze's reading of Whitehead is unquestionable. From *Difference and Repetition* onwards Deleuze takes up the principal directions of Wahl's thesis, *Pluralist Philosophies of England and America*. In *Dialogues*, Deleuze pays homage to Wahl, writing 'He [Wahl] not only introduced us to an encounter with English and American thought, but had the ability to make us think, in French, things which were very new; he on his own account took this art of the AND, this stammering of language in itself, this minoritarian use of language, the furthest' (Deleuze and Parnet, *Dialogues*, p. 58).
27. Wahl, *Vers le concret*, p. 153. (Translator's note: my own translation.)
28. Cited in PR, 25.
29. Leibniz, *Discourse on Metaphysics*, p. 47.
30. On the relations between expression and the expressed in Leibniz, see Deleuze, *Difference and Repetition*, p. 48.
31. Whitehead, *Science and the Modern World*, p. 73.
32. 'The passing condition, which involves and represents a multiplicity

in the unit [*unité*] or in the simple substance, is nothing but what is called *Perception*, which is to be distinguished from Apperception' (Leibniz, *Monadology*, p. 224).

33. For Whitehead, this explicitly involves questioning a definition of substance found, in particular, in Aristotle's *Organon*: 'A substance – that which is called a substance most strictly, primarily, and most of all – is that which is neither said of a subject nor in a subject' (Aristotle, 'Categories', p. 28), or '[i]t is a characteristic common to every substance not to be in a subject' (Aristotle, 'Categories', p. 31).
34. See Gérard Sondag's crucial introduction to Duns Scotus's *Le Principe d'individuation*, as well as Gilson's *Jean Duns Scot*, pp. 244–52.
35. Descartes, 'Meditations on First Philosophy', p. 75.
36. 'To say that a bit of matter has *simple location* means that, in expressing its spatio-temporal relations, it is adequate to state that it is where it is, in a definite finite region of space, and throughout a definite finite duration of time, apart from any essential reference of that bit of matter to other regions of space and to other durations of time' (Whitehead, *Science and the Modern World*, p. 58). Whitehead repeatedly characterises his project as a questioning of this principle: 'the element in this scheme which we should first criticise is the concept of *simple location*', or 'among the primary elements of nature as apprehended in our immediate experience, there is no element whatever which possesses this character of simple location' (Whitehead, *Science and the Modern World*, p. 58).
37. Translator's note: This is Whitehead's citation of the Haldane and Ross translation of the *Meditations* (Descartes, 'Meditations', in *Discourse on Method and Meditations*, p. 75, cited in PR, 41). For the equivalent passage in the Cottingham, Stoothoff and Murdoch translation, see Descartes, 'Meditations', in *The Philosophical Writings of Descartes*, p. 19.
38. Whitehead, *Science and the Modern World*, p. 75.

6

What is the Subject?

A New Economy of Object and Subject

In his reading of Descartes, Whitehead extracts a definition of the subject as a relation through which feelings are unified and appropriated. The key point of disagreement is found in the inverse relations that each constructs between the subject and feeling. If Whitehead does in fact take up the problem's terms, he is nevertheless radically opposed to the Cartesian economy organised around a subject *qua* foundation of feeling.

> Descartes in his own philosophy conceives the thinker as creating the occasional thought. The philosophy of organism inverts the order, and conceives the thought as a constituent operation in the creation of the occasional thinker. The thinker is the final end whereby there is the thought. In this inversion we have the final contrast between a philosophy of substance and a philosophy of organism. (PR, 151)

Whitehead's reading could be criticised, of course: he takes a Cartesian proposition, pushes it in the direction of speculative philosophy, only to return, finally, to Descartes' own internal coherence, opposing it to an entirely different economy of thought. This, however, would be to lose what is important in Whitehead's reading of Descartes. Whitehead is not doing history of philosophy. The relevance of each of his critiques and reprises could, of course, be justly attacked in so far as they are constructed on grounds that would have been completely foreign to the original thinkers. A highly singular relation to philosophy is set up in *Process and Reality*, a relation that deserves a whole work of its own. Whitehead proceeds by way of statements and propositions, attempting to evaluate both the internal effects of a philosophy as well as its speculative consequences beyond its inscription within a particular system. There can be little doubt, as Stengers points out in *Thinking with Whitehead*, that this practice of reading owes much to his past as a mathematician. When Whitehead writes that Descartes was an 'occasional thinker', his motivation is not so much to clarify an analytic point or to aspire to a kind of erudition about Cartesian philosophy as it is to give meaning to another, much more fundamental proposition, that thought is 'a constitutive operation in the creation of the occasional thinker' (PR, 151).

Either feelings come first and the subject (the actual entity) emerges out of these feelings, or the subject (the actual entity) comes first and feelings emanate from it as the consequences of the subject's own activity. This is the alternative Whitehead tries to dramatise, making it rest on an inverted Cartesianism: Whitehead takes this see-saw very seriously and works to make felt its full importance. His proposition is decisive because it is obvious a priori that the subject must in some way be given prior to its feelings. In other words, for the economy of subject and feeling to attain its true density it is necessary to affirm both the importance of the Cartesian proposition and the difficulty of breaking with it. And he goes further still: having borrowed the terminology from Descartes, he generalises the relation of subject and feeling in which he sees a mode of thought that traverses modernity.

> Thus for Kant the process whereby there is experience is a process from subjectivity to apparent objectivity. The philosophy of organism inverts this analysis, and explains the process as proceeding from objectivity to subjectivity, namely, from the objectivity, whereby the external world is a datum, to the subjectivity, whereby there is one individual experience. (PR, 156)

Kant activates a new terminology, a new vocabulary. According to Whitehead, however, the break is only superficial. The subject retains its primacy as a foundation. The question of the subject and feelings is nowhere to be found in Kant, but rather of the relation between subjectivity and objectivity, a relation which is at once very different and somewhat similar in the gestures it triggers.

The Subject-Superject Structure

In its present state, this reversal could give rise to a number of objections: how could objects be the cause of a subject if they were not, in some way, already animated by a subjective principle?[1] What could guide them towards the 'subjectivity' that Whitehead is talking about? These objections could be modified, of course, and yet, ultimately, they all come down to assuming that objects can give rise only to objects and, as such, where a *new* actual entity could come from cannot be seen. Ultimately, to speak of a prehension or an appropriation could seem like a simple displacement of the problem on to something else, leaving it otherwise unchanged. The whole problem, which can only be briefly outlined, is to know how a new entity can spring from old ones. And the objections to this possibility have to be taken entirely at face value. Indeed, disjunctive diversity, or, in the Kantian terminology taken up by Whitehead, 'objectivity', could not give rise to a new entity, 'subjectivity', if the latter did not in some way, in part, already exist.

How can Whitehead simultaneously accept the terms of these objections and refuse their consequences, the most important of which, and the most devastating from the perspective of the philosophy of individuation, would be the construction of a fully constituted subject, and, therefore, a breaking with the question of emergence and the demand of monism, both of which had directed him up to now? The answer is: by introducing a new distinction, a consequence of the difference between a constituted entity and an entity in the process of actualisation.

When we say that an entity is the subject of an individuation, the subject in question can be understood in two different ways. It can, first, be taken as the subject as classically conceived, emerging from the Latin tradition. This is the subject as '*subjectum*', meaning 'to be below', giving rise to the notion of subjectivity as it is found, especially, in Descartes. When Whitehead writes that the subject arises from its own feelings it is, for the most part, this *subjectum* that he means, the subject that gives the impression of being the foundation of its feelings. Such an impression comes from the fact that the *subjectum* emphasises self-identity, autonomy and an apparent independence from the environment in which it is located. It is a substantial subject, in the sense of the Cartesian notion of substance.[2]

This definition of the subject need not be refused: if it arises it is because it is effective at displaying particular qualities of existence, especially the real fullness of reality's constitutive elements. What damages it, and what Whitehead refuses to repeat, is a sort of exaggeration or radicalisation, the way it makes this definition the fundamental quality of the notion of the subject. *By identifying subject with* subjectum, *philosophy renders itself incapable of understanding genesis: the notion has been fabricated with the precise aim of expressing the fullness and autonomy of the subject.* How could it have expressed a subject's coming into existence when the very categories it puts in place to think the subject are directly opposed to all notions of transformation, augmentation, movement or change? The *subjectum* is that which by definition is insusceptible to either augmentation or diminution, or even intensification.

This is why a philosophy of the subject's emergence out of feelings is possible, though only on the condition that the relation of identification between the 'subject' and the *subjectum* is undone and another dimension is attributed to it, one that does not contradict it but limits its field of application. Again, this involves returning to the Latin source of the notion of the subject: the subject, after all, can also be thought as '*superjacio*', a term that can be translated by a number of expressions including 'to throw over', 'to throw towards', and 'to overtake' or 'to cross'. No longer a self-sufficient subject, adequate to itself, but a *stretched* subject, excessive in relation to its momentary identity. No longer identical to itself, it is projected beyond its factual existence. Whitehead translates this with 'superject'. And this concept of 'superject' can explain why a subject can indeed exist outside of categories of fullness, sufficiency and independence.

There is subjectivity in disjunctive diversity in so far as entities emerge or strive towards a fullness against which they are defined. One might say that 'feelings' are animated with virtual subjectivities that direct them towards something which as yet does not exist as such, given only as a tendency or as an 'aim'. The raised objections were correct: there can be feelings only when something *feels*, or integrates. Despite this, however, there is no need to posit the existence of a ready-made reality as the reason or foundation of feeling.

The distinction between the two meanings of 'subject' is a formal one: what belongs to the order of the 'subject' and what belongs to the order of the 'superject' cannot be clearly distinguished in any process of individuation. There is no difference in nature between the two, nor any difference based around moments of individuation: the tendency, the animating aim – what Whitehead also calls its 'principle of unrest' (PR, 28)[3] – acts within its concrete reality at every stage of the entity's emergence. As such, '[t]he feelings are inseparable from the end at which they aim; and this end is the feeler. The feelings aim at the feeler, as their final cause' (PR, 222). They strive towards the feeler, towards the subject, and yet they do so only in so far as the latter is presented as a virtual form of existence. Whitehead summarises this relation between the subject and the superject in the final section of the 'Categorial Scheme':

> An actual entity is at once the subject experiencing and the superject of its experiences. It is subject-superject, and neither half of this description can for a moment be lost sight of. The term 'subject' will be mostly employed when the actual entity is considered in respect to its own real internal constitution. But 'subject' is always to be construed as an abbreviation of 'subject-superject'. (PR, 29)

The Subjective Aim and Self-enjoyment

Although this distinction was constructed with the aim of diminishing the importance of the notion of 'subjectivity', it produces an unexpected consequence. Before attempting to explain it, we should ask: to what necessity does the notion of the subject, as identified entirely with 'subjectivity' (*subjectum*), respond? To impose 'subjectivity' as the only possible translation of the subject is to attempt to construct a theoretical 'bifurcation' of the real. In *The Concept of Nature* Whitehead gave considerable importance to the term 'bifurcation' in his attempt to analyse the theoretical foundations of modern science.[4] Although the context of the problem is not exactly the same, one could notice such an operation of bifurcation in the weight given to the notion of subjectivity. Of course, there is no longer any opposition between a real nature and an apparent nature, nature in itself and nature as it is perceived. The opposition, rather, is between that which depends on subjectivity and the qualities attributed to it, and a set of heterogeneous physical, chemical and biological realities that have nothing in common save for the contrast they establish with subjectivity. It follows, then, that the subject occupies a deeply peculiar position: it becomes the central though excessively limited element of experience across the set of operations that have extracted its qualities in order to send them back to the register of nature.

It would be difficult, in such a context, to speak of a subjectivity in nature, of the subjectivity of an atom, of a cell or a rock, since the concept was invented precisely to distinguish between orders of reality. It is this sharp separation that the Whiteheadian distinction between the two senses of the word 'subject' blurs and complexifies. The distinction with the original goal of limiting the importance of subjectivity will, by adding the dimension of the superject, considerably extend subjectivity's reach. Instead of being reserved for a particular order of reality, it becomes the essential element of actual entities, and so of existence in each of its modes. It is no longer absurd, then, to speak of the subject – in the sense of 'subjectivity' and 'superject' – with regard to realities as various as the atom and the organic body. These forms, as we will see, are not just identified with the notion of subjectivity but truly require it. Because they are supported by actual entities, these realities activate immanent subjectivities. On each plane the existential tensions between 'subject' and 'superject', efficiency and aim, actuality and tendency, are at work, engaged in the microscopic processes of individuation.

After all, terms like 'aim' and 'tendency' remain too imprecise to express the 'subject-superject' structure. The most adequate term remains 'subjective aim', which expresses the relation to what should be called the 'final cause' of actual entities, the driving cause of individuation. In classical terms, 'disjunctive diversity' is the *efficient cause* and the subjective aim the *final cause* of individuation.[5] There is, however, a strict condition for employing such a highly charged concept as 'final cause'. If by 'final cause' we understand a term which would in some way be pregiven, an essence to be realised or a form to be fulfilled, existing before the process of individuation itself, the ultimate principle – the emergence of novelty – would be contradicted. How could the ultimate principle be satisfied if we affirm, at the same time, that the entity aims towards a finality that already existed? The answer, here, is that an end does in fact exist, though it cannot be conceived outside of individuation itself. The end is immanent to each operation of prehension and capture. Individuation is not a sequence or a series, moving from disjunctive diversity towards an end through a succession of independent stages. Diversity remains active, directed by the 'subjective aim', and the latter is defined internally to prehensions. If by 'final cause' we understand aim as immanent, as the orientation of individuation, then the subject-superject structure is indeed the integration of final causes within the actual entity.

This immanent aim, actualised as individuation approaches its end, is called 'self-enjoyment'. It is very difficult to translate this expression into French. The impulse to translate it with *jouissance de soi* risks anthropomorphism, and would restrict the concept. Deleuze, again in *The Fold*, returns to this notion and locates it in a Neoplatonic tradition, bringing to light its fully cosmic, or at least non-anthropological, impact:

> self-enjoyment [. . .] marks the way by which the subject is filled with itself and attains a richer and richer private life, when prehension is filled with its own data. This is a biblical – and, too, a neo-Platonic – notion that English empiricism carried to its highest degree (notably with Samuel Butler). The plant sings of the glory of God, and while being filled all the more with itself it contemplates and intensely contracts the elements whence it proceeds. It feels in this prehension the *self-enjoyment* of its own becoming.[6]

Deleuze, here, accurately expresses the relation of intensity[7] in self-enjoyment: *a richer and richer private life*. One might say that individuation is indeed a relation of intensity: it means 'being filled with the world', prehending it, integrating it into its actuality. *The extension of prehensions leads to an increase in self-enjoyment.* The more an entity captures other entities, the more it experiences what belongs to it – the 'privacy' Deleuze speaks of – up to the 'ultimate' point where everything is prehended, where it is no longer exclusively a self-relation since everything is internal to it. It then becomes pure 'self-enjoyment', no longer a 'tendency', 'aim' or 'goal' but actual existence. There is, then, no opposition between others and the self at the level of actual entities, since every capture makes the entity's exclusive and unique identity more obvious. The subject, then, instead of implying a lockdown, a closure or an independence to the detriment of relations, is nothing but a pure relation, a multiplicity of integrated, and so internal, relations. An entity is both a singularity and a cosmic experience. Individuation can be illuminated by returning to the two meanings of the word subject in this way: intensive and relational, individuation amounts to a determination of relations from a point in experience, a perspective on everything that exists.

The Two Activities of Prehension: Exclusion and Feeling

When Whitehead says that the entity prehends everything, and that therefore disjunctive diversity is present with the new entity in its totality, this is not a provisional or metaphorical formulation: *everything is literally captured by the entity*, all potentials are mobilised, the total universe of actual entities 'conspires' in each new individuation. This poses serious problems for the scheme's coherence, of course. If every entity is indeed the prehension of all the others, then it is composed of all the others, which is to say that it is itself every actual entity. The question, then, is knowing how every entity can be at once singular and distinct while being composed of the same elements.

This question, again, is essentially technical. It is not about knowing if this or that prehension of everything that exists is true or false, if it is adequate or not to reality, if we agree or not with such a way of expressing existence. The question is entirely different: if prehension is asserted as the integration of everything, then we risk reducing or perhaps even contradicting the other constraints of the scheme's construction developed up to now. In so far as the problem is technical, its solution must be found in clarifications of the term 'prehension'. Everything is indeed prehended, but in two distinct modes. There are '(a) "positive prehensions" which are termed "feelings", and (b) "negative prehensions" which are said to "eliminate from feeling"' (PR, 23). The former are obvious and comprise the essential part of what we have analysed so far. But what is meant by prehensions which 'eliminate from feeling'? Returning to the literal meaning of 'prehension', we can speak of a capture, of an *appropriation of exclusion*. In leaving this contradiction behind, however, do we not merely produce another, slightly displaced one? Is there not a sense in which our wish to leave behind the contradiction of 'everything is integrated' simply introduces another contradiction, an idea of integration by way of elimination?

Let us take the hypothesis at face value and trace its consequences, starting with the principle that *eliminating something is not without effects*. If we reject particular elements in an action then either the elimination forces what is rejected into absolute exteriority, what is rejected leaves no trace behind, no mark on that which rejected it, relating elimination to indifference; *or*, on the other hand, something takes place in the action itself. In the *act* of rejection there is a relation that produces effects. The first option, in fact, is not really an option at all, since it requires translating elimination into simple indifference and, ultimately, denying the act of elimination itself, which is anything but indifferent. For Whitehead, the actions 'which effect the elimination are not merely negligible' (PR, 226). Something happens that cannot be reduced to the terms of elimination.

This is the fundamental point, to which 'prehension' and 'elimination' can be connected: *to reject or to refuse is, in one way or another, to build a link.*

> A feeling bears on itself the scars of its birth; it recollects as a subjective emotion its struggle for existence; it retains the impress of what it might have been, but is not. It is for this reason that what an actual entity has avoided as a datum for feeling may yet be an important part of its equipment. The actual cannot be reduced to mere matter of fact in divorce from the potential. (PR, 226–7)

This is what Peirce calls a 'would-be',[8] a possibility. What would or could have been, the choices made and the selections that took place, define an entity just as much as what it actually includes. Feeling carries along with it all those 'would-be's, those eventualities that had to be excluded from its effective existence. Everything takes place as if positive prehensions – effective feelings – were permanently accompanied by a constellation of negative prehensions, those avoidances, refusals and rejections which give actions their full importance. This is what Whitehead means when he writes that '[t]he actual cannot be reduced to mere matter of fact in divorce from the potential'; everything actual is overdetermined by a set of potentials, the actions which could have been integrated into the 'real constitution' of an entity but which had to be removed.[9]

Negative prehensions, however, although crucial, remain subject to positive prehensions. Only on the basis of the latter, joined to their side, do they have any real importance: 'negative prehensions which consist of exclusions from contribution to the concrescence can be treated in their subordination to the positive prehensions'

(PR, 220). Eventualities have consistency only by way of real affirmations of existence, true positivities. Besides, it is the very function of negative prehensions to bring out positive prehensions, to put them into perspective and to demonstrate their importance by drawing out the contrasts and the choices within possibilities. Rather than *relativising* real existence, the efficacy and actuality of a choice or an existential decision, then, the function of negative prehensions is precisely to mark out their *necessity*.

Monism prevents us thinking the difference between positive and negative prehensions as a difference between different forms of existence. Nothing a priori destines an actual entity to be rejected by any other, to be the object of negative prehensions. It is all a question – as it always is with actual entities – of economy and relations: in such and such a circumstance in relation to such and such an assemblage an entity is rejected; but in another, it is included. Here, then, we have the foundation of *a radical empiricism that definitively breaks with the question of the nature of the elements in question* in favour of pure relations and encounters. Incompatibilities between entities are connected to the complex relations emerging from the totality of relations that sustain the universe, to the extent that some other kind of relation might lead, on the contrary, to an inclusion. It is a question of harmonies.

> An actual entity has a perfectly definite bond with each item in the universe. This determinate bond is its prehension of that item. A negative prehension is the definite exclusion of that item from positive contribution to the subject's own real internal constitution. This doctrine involves the position that a negative prehension expresses a bond. (PR, 41)

We can now return to our question: what is a subject in *Process and Reality*? It is, first of all, a technical term, used to designate that which emerges as a novelty. In a process of individuation, the subject is what is individuated, it is what comes into existence. Secondly, this actualisation or emergence is achieved by way of captures or appropriations. The subject is a 'vector' which makes the 'there' pass into a 'here' by unifying the 'there' into a new form of existence (through the rejection of other real possibilities and inclusions). Finally, the 'subject' is what experiences itself *as its own*, a singularity filled with the whole universe of its prehensions, a self-enjoyment understood as the intensive experience of the sum total of its relations, an experience at once private and cosmic.

Notes

1. This criticism is not without analogy to Heidegger's response to the empiricists in which he takes up the idea of a 'turning toward': 'A finite creature must be able to take the being in stride, even if this being would be directly evident as something already at hand. Taking-in-stride, however, if it is to be possible, requires something on the order of a turning toward, and indeed not a random one, but one which makes possible in a preliminary way the encountering of the being. [. . .] The turning-toward must in itself be a preparatory bearing-in-mind of what is offerable in general' (Heidegger, *Kant and the Problem of Metaphysics*, p. 63).
2. 'By *substance* we can understand nothing other than a thing which exists in such a way as to depend on no other thing for its existence' (Descartes, 'Principles of Philosophy', p. 210).
3. Whitehead locates this principle within a tradition beginning with Samuel Alexander: 'every ultimate actuality embodies in its own essence what Alexander terms "a principle of unrest, namely, its becoming"' (PR, 28) And yet Alexander himself is only extending Leibniz's idea of 'imperceptible little urges' which 'keep us constantly in suspense', and which are 'confused stimuli, so that we often do not know what it is that we lack'. In these little urges Leibniz sees 'impulses' which are 'like so many little springs trying to unwind and so driving our machine along' (Leibniz, *New Essays*, p. 166).
4. 'What I am essentially protesting against is the bifurcation of nature into two systems of reality, which, in so far as they are real, are real in different senses. One reality would be the entities such as electrons which are the study of speculative physics. This would be the reality which is there for knowledge; although on this theory it is never known. For what is known is the other sort of reality, which is the byplay of the mind. Thus there would be two natures, one is the conjecture and the other is the dream' (Whitehead, *The Concept of Nature*, p. 30). For an analysis of the construction and the effects of the 'bifurcation of nature', see Cesselin, 'La Bifurcation de la Nature'.
5. The attempt to rehabilitate 'final causes' in a way that would complete rather than replace 'efficient causes' spans *Process and Reality*. Whitehead, speaking of Aristotle, writes: 'His philosophy led to a wild overstressing of the notion of "final causes" during the Christian

middle ages; and thence, by a reaction, to the correlative overstressing of the notion of "efficient causes" during the modern scientific period. One task of a sound metaphysics is to exhibit final and efficient causes in their proper relation to each other' (PR, 84).

6. Deleuze, *The Fold*, p. 78.
7. For an analysis of the concept of intensity in Whitehead, see the crucial work by Judith Jones, *Intensity: An Essay in Whiteheadian Ontology*.
8. Peirce, 'A Survey of Pragmatism', p. 319.
9. On the importance of negative prehensions, see Stengers, *Thinking with Whitehead*, pp. 212–13.

7

Realisation of Self and Power

The End of Individuation

The process of individuation has an end. The passage from disjunctive diversity to the unity of a new entity embodied by the subject has a conclusion, namely, the effective realisation of the entity, its full actualisation. This end point of individuation is reached following the determination of every positive and negative prehension of the entity, that is, when all of its relations with other entities have been established. It is, then, fully a perspective, a being-situated in the universe, a junction between and a unity of everything that exists. It attains, in its final state of concrescence, what Whitehead calls 'satisfaction'. This 'satisfaction' is not a common end, identifiable with all the others, as if there were a pre-existing finality in individuation that would be actualised in a particular manner. It is 'a generic term: there are specific differences between the "satisfactions" of different entities, including gradations of intensity' (PR, 84). In the same way that every prehension is singular and belongs to the subjective orientation of every actual entity, the end of an entity is specific, it is *that* end for *that* entity.

What is 'satisfaction' in *Process and Reality*? As with all of the concepts we have treated up to now, satisfaction is a 'generic' concept from which we can extract components, an extraction that does not reduce its form, which remains concrete and specific. Being in a state of 'satisfaction' (*satisfacio*) means, first of all, being 'settled', as one might say of a debt or an obligation. It has the juridical sense of a transaction in which one of the parties is acquitted of their contractual obligations, whether formal or informal. Transposed on to the plane of actual entities, it signals that the entity has been acquitted of its requirements and obligations

– particularly of its prehensions – and that it has responded to every actual entity that already exists. It has been acquitted of its obligation to take every existence into account. This is an ontological requirement, entailed by creativity as the ultimate: every existence is a unity of that which already existed. Second, satisfaction means being 'self-sufficient' [*satisfait*], as one might say of a thing or an individual when we are emphasising its 'autonomy'. The actual entity acquires a 'sufficiency' when it reaches its end, as if the fact of being fully determined gave it a sort of autonomy, a self-relation, a fullness of being.

These two meanings come together in the process of individuation. It is because the entity is freed from its obligation to be connected – its being has been identified with the relation that it can have with everything that exists – that it can attain a being in itself, a pure 'self-enjoyment' that is nothing other than the enjoyment of the particular mode of the relation it sustains with the universe. Whitehead summarises the meaning of this end of individuation with a triple determination:

> The final phase in the process of concrescence, constituting an actual entity, is one complex, fully determinate feeling. This final phase is termed the 'satisfaction'. It is fully determinate (a) as to its genesis, (b) as to its objective character for the transcendent creativity, and (c) as to its prehension – positive or negative – of every item in its universe. (PR, 25–6)

The first determination is that of its genesis, that is, of its own mode of becoming. This determination only repeats what we have already elaborated at length, namely, the identification of the entity's being with the process by which it comes into existence. The second determination, for its part, deserves a longer expansion. It follows from the first since it is a question of its being *after* individuation, its existence following the achievement of individuation. This becoming determinate, Whitehead tells us, now concerns its 'objective character', that is, its existence within another. This returns us to the distinction between 'formal' and 'objective' existence while clarifying its function: an entity's 'satisfaction' is the achievement of its 'objective character' for other individuations. And the third determination, finally, is the relational determination we have already analysed in the process of prehension.

This triple determination is the origin of the transformation of the *subject of individuation* into the *object of individuation*. As subject, it aspires towards its 'subjective aim', or, more precisely:

> In its self-creation the actual entity is guided by its ideal of itself as individual satisfaction and as transcendent creator. The enjoyment of this ideal is the 'subjective aim,' by reason of which the actual entity is a determinate process. (PR, 85)

And yet when it attains its satisfaction, its end, it is no longer 'guided' by this aim but is identified with it. Its being becomes its actuality. This is why Whitehead says that, in a certain way, 'satisfaction' is the *perishing* of the entity. What made it a living reality, after all, was a kind of disparity, namely, the subject-superject relation: the principle of agitation and becoming that pushed the entity into being something other than itself. It is a perishing in the highly precise sense of the end of a process. But this death is not a *disappearance*. The end of the process has absolutely nothing to do with the vanishing of the entity. It simply has to do with the fact that it is no longer the 'subject' of emergence and transformation. The perishing of the entity is the fact that its prehensions have all been produced, that its activity of capture and appropriation has reached an end in its effective realisation.

It does not disappear but rather comes to join what already exists: other entities. Like them, it becomes part of disjunctive diversity, present in the emergence of new actual entities. *Its end is the beginning of another form of existence*. Whitehead often returns to the Lockean formula that expresses a vision of one process of individuation following another, of an entity's end as the beginning of a new entity: 'time is a "perpetual perishing"'.[1] The *subjective life* of the entity disappears, but it now persists in its *objective life*, like 'our immediate actions, which perish and yet live for evermore' (PR, 351).

In this sense, in this new mode of existence, the entity acquires an 'objective immortality': 'actual entities "perpetually perish" subjectively, but are immortal objectively' (PR, 29). Immortality means that, from now on, the entity will be prehended by future entities, even if that means being eliminated by future negative prehensions. The individuation of actual entities, then, can be expressed in the following poetic formula to which Whitehead, in *Symbolism*, gave a great deal of importance: '*Pereunt et imputantur*' or '[t]he hours perish and are laid to account'.[2] 'Objective immortality' is nothing but this: the subject, having reached its end, is laid to account by new becomings.

An important consequence, then, is that 'becoming' and 'immortality' can no longer be separated.

> In the inescapable flux, there is something that abides; in the overwhelming permanence, there is an element that escapes into flux. Permanence can be snatched only out of flux; and the passing moment can find its adequate intensity only by its submission to permanence. (PR, 338)

Satisfaction is the exact point at which becoming is transformed into immortality, genesis into an existence given once and for all: 'The notion of "satisfaction" is the notion of the "entity as concrete" abstracted from the "process of concrescence"; it is the outcome separated from the process, thereby losing the actuality of the atomic entity, which is both process and outcome' (PR, 84).

As long as the entity displays agitation, as long as it is both effective and projected beyond itself as an 'aim' or a tendency, the question of its value and its meaning has no importance. That is, its meaning is precisely its 'subjective aim', its final cause immanent to its efficient causes. And yet, now that it is 'adequate' to itself, now that it is exactly what it is and nothing more, does it still have a meaning and a value? And if an entity still has meaning and value following its satisfaction, where is it situated? What expresses it?

Let us begin by imagining what would appear to be the most plausible option, that is, the meaning of an entity would belong to that entity, would be immanent to it. This would presuppose that the entity was capable, in one way or another, of expressing its satisfaction, of bringing it to light, whatever its form. And yet, logically, '[n]o actual entity can be conscious of its own satisfaction; for such knowledge would be a component in the process, and would thereby alter the satisfaction' (PR, 85). All consciousness, even in the most minimal form, every expression on any plane presupposes a movement or a change, a minimum of process. Satisfaction, however, is radical: an entity is no longer capable of the slightest change, it becomes an object without immanent becoming. As a result, to introduce meaning at the level of such an entity would end up initiating a new process, contradicting the very notion of its unchanging immortality. It is, then, impossible to find meaning within entities that have reached their end.

The second option remains: to locate the entity's meaning and value elsewhere, in another reality. Is there a domain, a sphere of existence, that has the role of expressing the meaning of each entity? To make such a claim would be to presuppose forms of existence other than those of actual entities themselves, contradicting the ontological principle. The only remaining option, then, is that their meaning is to be located in other processes of existence, that is, in other actual entities still in the process of individuation. The meaning of an entity is always within another. Whitehead speaks of a 'pragmatic value', not without relation to Peirce's principle analysed in the first section, namely, that the meaning of a statement is found in its effects. This is why the value of an entity is identified with its 'usefulness': 'the "satisfaction" of an entity can be discussed only in terms of the usefulness of that entity. It is a "qualification of creativity"' (PR, 85).

The 'satisfied' entity, through the simple fact of its existence, directs creativity, that is, it directs the possibility of the genesis of other entities that will include it, just as when it was still a prehending subject, it included other entities. A pragmatic approach is enforced, then, an approach in which meaning is located in objectification, in the encounter between the object and the subject, not in their analysis. No entity can express its own meaning; no entity can attempt to 'possess' or control it. The universe is one of pragmatic meanings, variable according to relations and harmonies; a universe of coherences produced by processes the novelty of which cannot be reduced to the contents of a prehension or to the meaning of any of its terms.

Power as the Direction of Potential

Satisfaction, however, does not imply passivity. The entity obviously loses the immanent activity of the 'subjective aim', but it acquires another form of activity. It passes not from an active to a passive mode but from *one form of activity to another*. This is why Whitehead introduces a concept into *Process and Reality* whose importance might appear to have been minimised by speculative philosophy's explicitly actualist orientation. This concept is the concept of *power*. When I defined the concept of 'actuality' above, I held that it was distinct from the being-in-power found in Aristotle. For the 'actualist' tradition it was imperative to avoid such a notion of 'power', which tended to reduce the importance of actual reality. Whitehead, however, reintroduces the concept, placing it at the centre of the relations between entities. *An entity is in relation with another not by external connections but by relations of power.* How does Whitehead maintain the actualist orientation of speculative philosophy while introducing such a concept of power, not only at the level of the relations between entities but also at the level of the entity's phases of existence?

In the case of the theory of feelings, Whitehead returned to Descartes' philosophy by deeply altering its original meaning. In the same way, here, Whitehead works with philosophers in whom he discovers an approach prefiguring his own construction of individuation. In this way he is able to clarify the constraints of the concept of power as well as its potential for inclusion in the speculative construction.

In *Adventures of Ideas* Whitehead links this question of power to the Platonic form. He writes: 'Plato says that it is the *definition* of being that it exert power and be subject to the exertion of power [. . .] It is the doctrine of Law as immanent.'[3] Whitehead finds in the *Timaeus* the possibility of identifying being – actual entities – with power. For Whitehead, Plato does indeed point out the problem and lays out its terms (being, power and immanence), and yet he remains at the level of 'suggestion' or 'general intuition'. This intuition matters because it establishes the direction that a metaphysics of power could take. And yet, when taken up in the mode of speculative philosophy, it is insufficient: it lacks an explicit systematisation of its conditions and implications. The latter, however, can be asserted in the frame of the general proposition that has, up to now, guided the entire construction of individuation: existence as actualisation.

Whitehead situates the essential elements of a speculative philosophy of power elsewhere. He takes from *An Essay Concerning Human Understanding* a way of posing the problem that gives priority to power over substantial being. The entire shift from the *Essay* to *Process and Reality* can be interpreted as the translation into speculative terms – with all of the transformations, modifications and displacements such a translation presupposes – of what the *Essay* developed in the context of a theory of the understanding.

Locke develops his concept of power in two distinct parts. The first is found in chapter twenty-one of the second book, exclusively concerned with power. The second is situated in chapter twenty-three of the same book, entitled 'Of our Complex Ideas of Substances'. These two occurrences are both found in the second book, entitled 'Ideas'. From this simple fact we can already deduce not only that power intervenes on the plane of *ideas* and their emergence and relations, in short, their mode of existence, but also that even substance itself, for Locke, has to be posed in the context of a theory of ideas. This already testifies to the *Essay*'s distance in relation to *Process and Reality* on the question of power: for Locke it is a question of a problem connected to ideas. For Whitehead, however, it is necessarily a problem pertaining to existence.

We can begin, then, with the first occurrence of the notion of power in the *Essay*: chapter twenty-one. I have said that power intervenes on the plane of ideas. The question to ask, then, is what is the power of an idea? Locke's response focuses on the relation between the mind and ideas.

> The mind [. . .] considers in one thing the possibility of having any of its simple *Ideas* changed, and in another the possibility of making that change; and so comes by that *Idea* which we call *Power*. [. . .] Power thus considered is twofold, *viz.* as able to make, or able to receive any change: The one may be called *Active*, and the other *Passive Power*.[4]

This definition of power given by Locke is a quasi-literal take up of Aristotle, when, in book Θ of the *Metaphysics*, he connects it to change.[5] The definition of power in this passage can be summarised as follows: it is either the capacity for an idea to be changed or the capacity to make a change. Power, then, is divided, as if we could distinguish between two types of power: *to produce or to receive* a change. But we should not be mistaken, here: distinguishing them means only that power, in its unity, is expressed as either producing or as receiving a change. It is the *capacity* of being either one or the other that makes power. Locke clarifies the terms of this way of expressing power: when it is productive of a change it is called 'active power', and when it is the receiver of a change it is termed 'passive power'. Power is the expression of *capacity*, that is, of *possible activity*. Locke turns it into a principle: all power relates to action.[6]

It would be an error to identify 'passive power' with a kind of passivity: what is expressed is always an action or a change. The difference between the two powers is only in the *relation* between the terms: one creates and the other receives a change. Locke offers an example: 'Fire has a *power* to melt Gold, *i.e.* to destroy the consistency of its insensible parts, and consequently its hardness, and make it fluid.'[7] As such a moment, in relation with gold, fire is the active power. Fire produces a change in the gold by destroying the 'consistency of its insensible parts', the visible effect of which is fluidity. And yet fire would have no power if gold were not already disposed to accept this change, to be made fluid. The active power of fire is active only by encountering the passive power of gold, which is thereby appropriated and transformed. For Locke, then, the reasons for the change produced by their encounter could not be found through an analysis of fire or gold alone. Here, then, the question of natures is opposed by an empiricism of encounters: *such and such a power encounters another*. Fire can indeed produce a multiplicity of distinct effects according to the body it encounters: it can melt gold, it can blanch wax,[8] and so on. In the encounter of two ideas, then, what matters is *the relation itself, not the terms*. And one could say that there is not one single idea in the understanding that does not alternate between active and passive power. Since every idea can be connected to all others[9] they can be 'passive' or 'active' only relative to the established relation; and sometimes, if the relation involves more than two terms, they can be both at the same time.

An asymmetry persists, however, in the relation between active and passive powers. The two are interdependent, of course: there are no passive powers without the active powers that give rise to them and vice versa. Active powers, however, unlike passive powers, are *indifferent* to their effects. The sun, for instance, is indifferent to its melting of the wax – it simply expresses its own being. It is the wax that turns the sun into an active power, and does so to the degree to which it is changed. As for the sun, it is limited to provoking changes relative to the bodies it encounters.[10]

This brings us to the second occurrence of the concept of power in the *Essay*. Locke does not treat the idea of power directly but rather that of 'substance'. In this way he activates what he developed in his previous sections, treated above. To the question *what is a substance and where does our idea of it come from?* Locke responds that we should not be surprised 'that *Powers make a great part of our complex* Ideas *of Substances*'.[11] Instead of being a primary concept, the notion of substance is derived; it is a 'complex idea', an idea produced by the understanding. And power plays a major role in the production of this idea. Because we are used to seeing powers in ideas we form an idea of substance as the expression of those powers. What we actually form, however, is the illusion of 'something' underlying ideas that would be their origin. What we have, then, are not substances but *ideas and the relations* between them, which is to say, more importantly, *relations of powers*.

Locke quite clearly characterises his project in relation to power: 'my present Business being not to search into the original of Power, but how we come by the *Idea* of it'.[12] This means that the question of power needs to be situated at the level of its functioning in the mind, not at the level of an explanation of the real. Hume will take up the reduction of power to the mind in a more radical manner when he writes in the *Treatise* that

> I believe the most general and most popular explication of this matter, is to say, that finding from experience, that there are several new productions in matter, such as the motions and variations of body, and concluding that there must somewhere be a power capable of producing them, we arrive at last by this reasoning at the idea of power and efficacy.[13]

If Whitehead recognises Locke in so far as he gives an essential function to power, he refuses to turn that function into the expression of ideas and their relations in the understanding. What, then, does he take from the *Essay*? What transformations does power undergo when the problem is shifted from a theory of the understanding to a theory of existence? Let us look again at the generic principles Locke sets down on the subject of power: 1) an active power is *an idea's capacity for inciting* a certain effect in another; 2) active powers are *indifferent* to their effects; 3) every active power is linked to a *multiplicity of effects* according to the passive powers upon which they bear. These are the three central propositions that Whitehead will take up, almost as they are, to describe the functioning and the relations of actual entities. Immediately, then, he draws a line between himself and Locke.

> Locke, throughout his Essay, rightly insists that the chief ingredient in the notion of 'substance' is the notion of 'power'. The philosophy of organism holds that, in order to understand 'power', we must have

> a correct notion of how each individual actual entity contributes to the datum from which its successors arise and to which they must conform. (PR, 56)

The actual entity that 'contributes to the datum' is the equivalent of what Locke calls an active power. Just as the sun is indifferent to its effect on the wax, once the actual entity has attained satisfaction it is indifferent to its effects. The actual entity becomes a constraint imposed upon new entities simply through the fact of its existence. In one way or another new entities will have to take into account and even to 'conform to' the directions imposed upon them by its existence. This active power, then, is what we have analysed under the name of *objectification*, that is, the passage from the formal being of an entity to its objective being: 'the "power" of one actual entity on the other is simply how the former is objectified in the constitution of the other' (PR, 58). And since Whitehead calls 'potential' those entities which compose objective diversity, it could be said that we are dealing, here, with *the power of the potential in an individuation*, the power to direct and specify such an individuation.

These 'givens' of diversity would be incapable of producing a new entity if the latter did not appropriate them in a specific way. Prehensions are the equivalent of Locke's 'passive powers', fully dependent on 'active powers' without being reducible to them. It should be said, then, that the active power of disjunctive diversity encounters the passive power of the actual entity's prehensions. And yet they are as much two complementary expressions of two states as they are of one single entity: as long as the entity is the subject of becoming it is a passive power, and as soon as it attains its satisfaction it becomes an active power. Individuation, then, can be qualified as the perpetual encounter between two powers – diversity and the actual entity – or as the passage from one power to the other – subject and object.

Notes

1. 'The ultimate evil in the temporal world is deeper than any specific evil. It lies in the fact that the past fades, that time is a "perpetual perishing"' (PR, 340).
2. Whitehead's commentary on this formulation is interesting, although it mobilises two central notions in the analysis of perception (efficient

causality and presentational immediacy): 'Here "Pereunt" refers to the world disclosed in immediate presentation, gay with a thousand tints, passing, and intrinsically meaningless [. . .] "Imputantur" refers to the world disclosed in its causal efficacy, where each event infects the ages to come, for good or for evil, with its own individuality. Almost all pathos includes a reference to lapse of time' (Whitehead, *Symbolism*, p. 56).

3. Whitehead, *Adventures of Ideas*, p. 120.
4. Locke, *Essay*, pp. 233–4.
5. Power, in this highly particular definition, is 'a principle of change in another thing or in the thing itself regarded as other' (Aristotle, 'Metaphysics', p. 3563). There is no doubt that this definition is much less used than the other definition Aristotle gives of power in the same book: the power of being, power as the capacity to be something. These two meanings of power – capacity to produce a change and capacity to be something – have given rise, when they were translated into Latin, to the concepts of *potentia* for the former and *possibilitas* for the latter. See Heidegger's commentary: Heidegger, *Aristotle's Metaphysics Θ 1–3*.
6. See Locke, *Essay*, p. 235.
7. Locke, *Essay*, p. 233.
8. Locke, *Essay*, p. 233.
9. '[T]here is *no one thing*, whether simple *Idea*, Substance, Mode, or Relation, or Name of either of them, *which is not capable of almost an infinite number of* Considerations, in reference to other things' (Locke, *Essay*, p. 321).
10. Leibniz, in the *New Essays*, criticises Locke's conception of power. For Leibniz, '[s]ome have pushed this doctrine so far that they have tried to persuade us that if someone could touch the sun he would find no heat in it' (*New Essays*, p. 133). Radicalising the difference between powers, it becomes impossible to understand why what a particular power produces on a certain body has something in common with what it can produce on another body. As such, Leibniz continues, 'I would venture to say that if the melted or blanched wax were sentient, it too would feel something like what we feel when the sun warms us, and it would say if it could that the sun is hot. This is not because the wax's whiteness resembles the sun [. . .] but because at that time there are motions in the wax which have a relationship with the motions in the sun which cause them. There could be some other cause for the wax's whiteness, but not for the motions which it has undergone in receiving whiteness from the

sun' (*New Essays*, p. 133). The relations in the internal movements of the wax are what connect it to this particular power of the sun, and although the latter can produce similar effects in other powers, it nevertheless remains specific. This is precisely how Whitehead criticises Locke. Whitehead, however, unlike Leibniz, attempts to protect the exteriority of powers at the same time as maintaining their relations. This is what allows him to situate the problem in the philosophy of individuation. I return to this below.

11. Locke, *Essay*, p. 300.
12. Locke, *Essay*, p. 234
13. Hume, *Treatise*, p. 106.

8

Pure Potentiality and Actuality

What is an 'Eternal Object'?

In the chapter titled 'The Categorial Scheme', Whitehead introduces a new distinction, doubtless the most important in *Process and Reality* as well as the most difficult to understand: actual entities are distinct from what he terms 'eternal objects':

> the fundamental types of entities are actual entities, and eternal objects; and [. . .] the other types of entities only express how all entities of the two fundamental types are in community with each other, in the actual world. (PR, 25)

This unequivocally expresses the importance of eternal objects. They are placed on the same plane as actual entities, with which they share a common characteristic of being the 'fundamental types of entities': 'actual entities and eternal objects stand out with a certain extreme finality' (PR, 22). We are now at a point where there can be no doubt as to the place occupied by actual entities in *Process and Reality*. They form the centrepiece of the entire speculative construction. When Whitehead writes in the preface that the 'positive doctrine of these lectures is concerned with the becoming, the being, and the relatedness of "actual entities"',[1] it is now clear that he was literally expressing the project of the book. All of the elements set in place so far have concerned actual entities. These elements were entirely involved in the mode of existence and the emergence of actual entities. When Whitehead writes, then, that the two fundamental types of entities are actual entities and eternal objects, this is no neutral distinction, a simple invention of a new term; rather he presents us with an important problem, the consequences of which for the system are on a different scale to the set of elements considered so far.

Let us return to the passage. What is the substance of Whitehead's distinction between actual entities and eternal objects? It would be tempting to read it as a distinction between two separate spheres of reality, as if one – eternal objects – had been added to or stuck on to the other, as if a new world with its own particular existences had been set up, in parallel to the world of beings in action that have mattered up to now. After all the reservations and care taken to stay true to *Process and Reality*'s fundamental principles, namely that of 'ontological monism' and the 'ontological principle', Whitehead appears to reintroduce a radical dualism, suddenly and without explanation: two spheres from which everything else would be derived. This idea is all the stronger for the fact that Whitehead, in his previous works, especially in *The Concept of Nature*, did indeed construct a dualism between 'objects' and 'events', a distinction that he appears to repeat by differentiating between 'eternal objects' and 'actual entities'.[2] In *The Concept of Nature* he was committed to an ontological dualism in which two types of fundamental entities exist and interact in each concrete event.

Numerous commentators, especially in France,[3] who were interested above all in *The Concept of Nature*, saw *Process and Reality* as an extension and metaphysical generalisation of problems raised by the former work. With greater and lesser degrees of subtlety, they took up the difference found in *The Concept of Nature*, as well as the terms upon which that difference is constructed, and applied it to *Process and Reality*. Once performed, the implications of this separation for concrete existence, its relations and interweavings, can be analysed. And yet this division gives rise to a genuine incoherence in the speculative scheme, and, above all, creates an obstacle that is difficult to overcome if individuation is to be thought as the production of new existences. The risks of a Platonic reading, in the most classical sense of that term, that is, a participation of the sensible (actual entities) in the intelligible (eternal objects), are clear.

We can pose the question differently: what is the function of eternal objects in *Process and Reality*? To what necessity does their introduction into speculative philosophy correspond? To answer, we should consider the effects of the inverse idea, that is, we should evaluate the consequences for *Process and Reality* if Whitehead had never introduced 'eternal objects'. There would be, as shown up to now, only actual entities and their relations. Prehensions would be exclusively 'physical',[4] that is to say, they would be unifications of other actual entities. If it were the case that 'physical prehensions' were radicalised to the point of being the exclusive forms of prehensions, it would follow that the 'new' entity would be nothing but the return of everything that went before it, that is, disjunctive diversity. Its individuation would be nothing other than the repetition of what includes it, of all the other actual entities that have already attained their satisfaction. Its existence would be entirely determined by efficient causes, by its inheritance, its repetition of everything that exists. Where, then, would novelty come from? How could this singular and irreducible existence come into being, the existence that the ultimate principle places at the centre of the entire philosophy of individuation? If it were merely a return, a repetition, how could it be 'new'?

We have not stopped reiterating this idea in different modes: although an entity is nothing other than an 'inheritance' or a 'repetition', it is a repetition in a certain mode, in a particular manner that cannot be reduced to what it repeats. An entity is only, exclusively, a prehension. And yet the way in which it prehends does not emerge out of its prehensions. This 'manner' is the origin and source of novelty. Whitehead expresses it with expressions like 'manner', 'mode' or 'how'. To put it in the terms of power, it could be said that active powers constitute entities, but they do not explain and do not establish their passive power, their capacity to appropriate and to integrate. Without the hypothesis of 'eternal objects' we would find ourselves in a mechanistic universe, a universe in which novelty would be reducible to previously existing causes. With the concepts already developed it would, of course, be a peculiar sort of mechanistic universe: it would be made of individuations, of prehensions, of captures, of objective immortalities. It would, however, be determined by the fundamental idea that 'novelty', in one way or another, is explicable by the 'old', by the 'already existing'. The final causes of individuation, then, would be mere appearances in relation to the real and explanatory efficacy of 'efficient causes'.

When 'eternal objects' are introduced into such questions of repetition and novelty, then, it seems as if they have been there all along, unnamed, throughout the whole construction of the philosophy of individuation, from disjunctive diversity to satisfaction. The function of 'eternal objects' becomes obvious when they are introduced into this constellation of problems. They determine the 'how': how do actual entities include others? How do they realise themselves? And, finally: how do they inherit and how do they leave behind an inheritance? This question of the 'how' is at the heart of what, by taking an expression from Deleuze, we have termed a 'certain singularity'. In this expression it is clear that 'eternal objects' are anything but 'objects'. They have no form, no mode of existence. They refer, rather, to abstract forms, apart from our modes of perception: *sensa* such as 'green' and 'blue' but also the nuances of colours. Eternal objects are the universals of quality, they are *sensa* like redness that are 'felt with emotional enjoyment' (PR, 315). They are qualities of form and intensity. They are states or convictions, like 'being loved' or 'joy'. They also include objects of an objective type such as mathematical forms. They are named by terms like 'each' and 'exactly that'. They are 'patterns' and 'relations'.[5]

This is why Whitehead, aiming to distance himself from the ambiguity connected to the term 'object', calls them 'determinants'. A 'how' is not an 'object', it is a determinant of prehension. I am in full agreement with Ivor Leclerc when he writes that: 'thus in so far as eternal objects enter into this prehension, it is not as themselves data, but only as determinants of the definiteness of the data'.[6] The expression 'determinants of definiteness' might seem even more obscure than 'objects'. It does seem, however, to be more adequate to the problem, removing any ambiguity regarding its relation to previous works such as *The Concept of Nature*. Technically, we should speak of 'eternal determinants', of the determinants of prehension and individuation. This, then, is another advantage of the notion of 'determinants', namely, that it directly connects the function of 'eternal objects' to actual entities. In saying 'object' we risk visualising a fully constituted existence. By speaking of 'determinants', however, their relation to the entities is necessarily emphasised: we speak of determinants *of* something or *for* something.

The reason why the determinants cannot be reduced to the 'data' of the prehension is now clear, but to what necessity does the idea of such determinants as 'eternal' respond? Why does Whitehead give them this first and essential quality that he denies to every existence, namely, eternity? Again, the answer is purely technical: it is to be found within the constraints of the system, not in some postulate that would assert additional dimensions of reality. Whitehead did not attempt to introduce this notion of eternity at any cost. It is, rather, the logical consequence of their distinction from actual entities. We can return to some propositions: 1) the concept of the actual entity is identified with existence; 2) existence is essentially a becoming and, reciprocally, all becoming is located within an actual entity; 3) the notion of becoming means genesis, the coming into existence of an actual entity. It follows, then, that all genesis is uniquely that of actual entities. As such, *if eternal objects were the subjects of a becoming, if they could come into existence, then they would come from actual entities, that is, from concrete becomings.* The answer, then, is astonishingly simple: if we accepted that eternity were nothing but a quality shared with the notion of becoming then we would be imitating the very same operation of reduction that Whitehead denounces, namely, basing 'determinants' upon 'data'. Eternity refers to the regime of being where there is no birth, origin, transformation or end, all notions that have been ascribed exclusively to the level of actual entities.

For this reason, eternal objects are 'potentialities', pure possibilities; they refer directly to nothing that exists. Remember that 'potentiality' was already used to refer to 'disjunctive diversity' as and when the latter is involved in processes of individuation. In the same way, we can say that 'eternal objects', the 'determinants' of individuation, are 'potentialities'. Clearly, however, they are not potentialities in the same way, which is why Whitehead distinguishes them: 'eternal objects' form what he terms 'pure potentiality', whereas 'disjunctive diversity' constitutes the 'real potentiality' of individuation. The entire difference between these two forms of potentialities, both active in the genesis of actual entities, is found in the fact that the 'real' is defined as 'being in action' whereas the 'pure' is an abstraction, independent in relation to every determined existence. This new dimension of pure and abstract possibility allows the necessity of eternal objects to be clarified. The question, however, remains the same: how does the form of dualism that Whitehead introduces conform to *Process and Reality*'s monistic demand? What now has to be analysed are these relations between dualism and monism, between the difference of actual entities and eternal objects and the demand that reality consist of a single plane.

Transforming Platonism: Ingression and Participation

This question refers back, in a complicated way, complete with displacements and retranslations, to Plato's *Timaeus*. References to Plato are everywhere in *Process and Reality*. Seen alongside references to Descartes, Locke or indeed Leibniz, this could be taken as a kind of eclecticism on Whitehead's part. And yet his gestures of incorporation respond to particular necessities. Whitehead does not hesitate to place the speculative project into a Platonic lineage: 'the train of thought in these lectures is Platonic' (PR, 39). The proximity of *Process and Reality* to Plato's *Timaeus*, so openly affirmed in this passage, might be surprising. In the reading proposed so far, after all, have we not experienced a distinct form of anti-Platonism at work in *Process and Reality*'s assertions that what matters are singular existences, feelings, becomings, prehensions, and so on?

Whitehead is well aware of this discrepancy. He attributes it, however, to a history of readings of Plato, a distortion effected by interpretations that have tried to systematise general intuitions. By the name Plato, Whitehead writes,

> I do not mean the systematic scheme of thought which scholars have doubtfully extracted from his writings. I allude to the wealth of general ideas scattered through them. (PR, 39)

There can be no doubt, however, that the difference Whitehead marks out between a Platonism of 'general ideas' and that of a 'systematic scheme of thought' is addressed less to specialists, to readers of Plato, than it is to Plato himself. Whitehead aims not to recover some kind of purified and original 'Platonism' but to introduce elements into Platonic thought to effect a rupture, though in the shape of a return. In the same way that a reversal of Kantianism might be termed 'Kantian' (the objective becomes the precondition of the subjective, a sort of Copernican counter-revolution), Whitehead asserts a Platonic inheritance by completely inverting the theory of ideas.

What is this purified Platonism that Whitehead feels he has located in the *Timaeus*? It is, above all else, a dualistic Platonism with no hierarchy between forms of existence. It is a universe composed, first, out of

> that which keeps its own form unchangingly, which has not been brought into being and is not destroyed, which neither receives into itself anything else from anywhere else, nor itself enters into anything else anywhere [. . .] invisible – it cannot be perceived by the senses at all [. . .][7]

It is precisely this order of 'ideas' or 'forms' that Whitehead tries to renew with the concept of 'eternal objects':

> I use the phrase 'eternal object' for what in the preceding paragraph of this section I have termed a 'Platonic form.' Any entity whose conceptual recognition does not involve a necessary reference to any definite actual entities of the temporal world is called an 'eternal object'. (PR, 44)

Secondary to these forms is

> that which shares the other's name and resembles it. This thing can be perceived by the senses, and it has been begotten. It is constantly borne along, now coming to be in a certain place and then perishing out of it [. . .][8]

It is this order of the sensible that Whitehead connects to actual entities. What interests Whitehead in the *Timaeus* is that these two orders seem irreducible. Their interaction occurs only in the 'receptacle', a given place in which ideas are actualised, are differentiated and participate in spatio-temporal events.

This extremely stripped-down reading of the *Timaeus* has only one condition, namely, that the theory of participation is ignored. The simple fact that Plato writes that the subjects of the sensible share the same name as the objects before them and that they are 'similar' to them indeed indicates that there is a correspondence between the two orders in the form of a *resemblance*, to the detriment of sensible reality. *Platonic dualism disappears in the order of resemblance established between the two regimes of existence*. Whitehead, of course, does not overlook this reduction, though he attributes it to an exaggeration that Plato shares with the most important forms of Greek thought: an excessive valorisation of the mathematical model.[9] This exaggeration of the importance of deduction in mathematics and philosophy's generalising repetition of it led Plato to reduce the importance of efficacy, of action, in favour of pure possibility. As Christian writes:

> One way of putting Whitehead's objection is to say that this platonic view absorbs actuality into possibility. Actuality evaporates from the world of things. This world becomes infected with unreality. Whitehead holds on the contrary that nothing is more real than temporal individuals.[10]

This amounts to introducing, in *Process and Reality*, a profound transformation of Platonism in the guise of an inheritance. Transposed into the principles of a speculative philosophy, the meaning and importance of deductive models are limited. 'What matters, already at this stage, is not to attribute to [eternal objects] the responsibility that Plato attributes to his Ideas.'[11]

Continuing this focus on the significance of the possible, the importance of action and efficacy, Whitehead opposes another operation that he calls 'ingression' to the model of participation – a model that attempted to find solutions to the correlation of the sensible and the intelligible. Thanks to this concept of ingression, Whitehead can affirm a radical dualism between 'actual entities' and 'eternal objects' at the same time as an ontological monism. The analogy with Plato is clear, since Plato also sought to integrate dualism into a monism of ideas.[12] Whereas for Plato this entailed an *operation of reduction*, however, for Whitehead it becomes an *operation of communication* within one single register of being. What is ingression?

> The term 'ingression' refers to the particular mode in which the potentiality of an eternal object is realized in a particular actual entity, contributing to the definiteness of that actual entity. (PR, 23)

Ingression, then, is the process by which potentialities are inserted into actual existences, becoming '"ingredients" in actual entities'.[13] The relation between 'potentiality' and 'actuality', then, is reversed: actual entities do not find their existence or their real being in potentiality, but potentiality nevertheless has its own being, finds its existence in actual entities, in its ingression. Wahl, quite rightly, sees in this the centrepiece of a particular form of empiricism: 'this is the foundation of empiricism: eternal objects [. . .] tell us nothing about their ingression into experience.

To see this ingression requires venturing into the domain of experience.'[14]

No 'eternal object' can determine the forms of its ingression a priori: the actualisations of eternal objects are not contained within entities. And if this empiricist hypothesis is unfolded, a hypothesis that forces us to experience ingression as a singularity in perpetual repetition, then we arrive at another idea necessary to maintain the demands of creativity: *no particular ingression of an 'eternal object' can fully express what it is*. We do not know what a 'pure potentiality' can do a priori. Even the experience we might have of it, always local and particular, never allows us to say what a particular 'eternal object' is outside of the factual mode of its occurrence, its indication of an embodied possibility.

> The eternal object is exactly what it is, and as such, it is unknowable, unnamable. This is so, not in the manner of the God of negative theology, by his eminence, but because the verbs 'to know' or 'to name' refer to (sophisticated) modes of feeling, all of which presuppose the determination of the 'how'. The idea of describing an eternal object, even by analogy, is as hopeless as the idea of seeing the eye seeing, or understanding the mind understanding.[15]

Whitehead's actualist empiricism opens on to an empiricism of pure potentialities, infinite in their actualisations, always unknowable as they are but implicated in the entirety of factual existence. The dualism is maintained – action and pure potentiality are laid out in their radical heterogeneity – and yet such a dualism is thought internally to a monism in which every potentiality is connected to an actuality. Every existence is situated within a being-in-action that never exhausts that existence's own forms while conditioning the existence of those forms.

The Mode of Reality of Eternal Objects

'Eternal objects', then, are not objects of knowledge. This contrasts them with the Platonic idea, characterised by Plato as 'invisible – it cannot be perceived by the senses at all – and it is the role of understanding to study it'.[16] There is no faculty or point of view in experience that would allow one to say or to experience what an 'eternal object' is. The constraint of empiricism is severe, here: *we can experience what an eternal object can do only through ingres-*

sion. Some of the concept's components, however, do emerge out of a contrast with 'actual entities', and they can be outlined in the form of three general propositions. A systematic exposition of these components is nowhere to be found in *Process and Reality*, but it is possible to reconstruct them – with all the requisite cautiousness whenever Whitehead's philosophy is concerned – by relating them to particular occurrences of the problem, as well as to other works.

1. *Eternal objects are abstract by nature*. This is the logical consequence of the manner in which 'eternal objects' are qualified in *Process and Reality*, namely, as 'determinants' or as 'pure potentialities'. It is not *by accident* that they are characterised as abstract, which would assume some operation of abstraction by which 'eternal objects' would be 'released' in some way from concrete existences. On the contrary, they are *essentially* abstract in their nature. They exist only through their ingressions, without ever being entirely adequate to or identifiable with them. Not only do eternal objects give away nothing as to the form their ingression might take, but their ingressions reveal nothing about the nature of eternal objects. This is a complex relation between immanence and transcendence. At the same time as they are fully engaged in existence, they are indifferent or neutral in relation to it. It is crucial to maintain their 'ontological neutrality', their surveying [*être-en-survol*][17] of potentiality that enables them to *simultaneously be incarnated in several entities without altering their nature*, without changing or modifying in relation to their actualisations.

> An eternal object is always a potentiality for actual entities; but in itself, as conceptually felt, it is neutral as to the fact of its physical ingression in any particular actual entity of the temporal world. (PR, 44)

2. *There are no new eternal objects*. Essentially, this characteristic is simply the logical outcome of the affirmation of the eternity of eternal objects, analysed above. What is eternal, by definition, has neither beginning nor end. To speak of 'new' eternal objects, then, would presuppose a moment of inexistence, an emergence that would contradict the very nature of these objects. This refusal, however, can be completed with another, clarifying its scope and once again associating Whitehead with Leibniz. In *The*

New Essays Leibniz writes that 'there are ideas and principles which do not reach us through the senses, and which we find in ourselves without having formed them, though the senses bring to our awareness'.[18] The senses, as actual givens, are the 'occasions' of ideas. As the determinants of individuation, they cannot be produced by individuation. If they were, we would have to presuppose a moment within individuation at which an eternal object was not already in a relation of ingression. Given the function of eternal objects in prehensions, this would be impossible.

3. *Eternal objects have relational dimensions.* If we can conceive that 'each eternal object is an individual which, in its own peculiar fashion, is what it is',[19] just as actual entities are, we cannot, in concrete situations, abstract them from their relations with all the others. There is a relational universe of eternal objects, a universe of connections, of systems varying continuously according to their ingressions into actual entities. As such, '[a]n eternal object, considered as an abstract entity, cannot be divorced from its reference to other eternal objects'.[20] Its individuality is inscribed in a relational space. Whitehead describes two levels of such interlacing in *Science and the Modern World*:

> In other words: If A be an eternal object, then what A is in itself involves A's status in the universe, and A cannot be divorced from this status. In the essence of A there stands a determinateness as to the relationships of A to other eternal objects, and an indeterminateness as to the relationships of A to actual occasions.[21]

We find a similar approach in the work of the mathematician Albert Lautman, who also claims a Platonic lineage when he writes, in his *Nouvelles recherches sur la structure dialectique des mathématiques*, of a dialectic of ideas that would be a dynamism of 'pure potentialities', that is, of varying relations between different elements of potentiality. He writes: '[w]e have tried to show [. . .] how the ideal relations of an abstract dialectic, beyond mathematics, is realised in mathematics'.[22]

Lautman's problem is that of mathematics and its foundation. His question is of *the construction of a problem and the ever local solutions that the problem can receive.* Whitehead's problem, however, is ontological: how can a multiplicity of pure potentialities arrange and co-determine itself in concrete existence? Both, however, converge around the idea that there would be an infinite set of pure potentialities engaged in dynamic relations,[23] conditioned and directed by the existence of the actual world.

Notes

1. Whitehead, *Process and Reality*, p. xiii.
2. The concept of 'eternal objects' in Whitehead's work has a long history. In *The Concept of Nature*, the question of 'objects' emerges from a very particular problem, namely, the experience of nature in perception. This limit has been indicated above. *The Concept of Nature* never aims to leave perception, and the specific experience of it, behind. In his analysis of perceptual experience, Whitehead distinguishes between two heterogeneous faculties. The first he names 'sense-awareness', which is a particular kind of sense, a sensitivity to events as the primary 'matter' of perception. These events have three characteristics: 1) they are passages or durations and, as such, '[w]hatever passes is an event' (*The Concept of Nature*, p. 124). 2) They are essentially singular and heterogeneous: one could certainly 'observe another event of analogous character, but the actual chunk of the life of nature is inseparable from its unique occurrence' (*The Concept of Nature*, p. 169). 3) Events have an *extension*, which is to say that every event is composed of parts and is itself part of larger events. As such, '[t]here is the part which is the life of all nature within a room, and there is the part which is the life of all nature within a table in the room. These parts are limited events' (*The Concept of Nature*, p. 75). These 'events' are only a part of our experience of the 'passage of nature'. They are key factors of our perception, and yet they require another dimension of perception that Whitehead calls 'sense-recognition'. Here Whitehead introduces the notion of 'object': objects are the factors of the perceptual experience of nature that are 'repeated', are found in different places, in different events. 'Objects', then, are constructed through a contrast with 'events': 'Whatever passes is an event. But we find entities in nature which do not pass; namely we recognise samenesses in nature' (*The Concept of Nature*, p. 124). Examples of objects are *sensa*, such as 'colour' (*The Concept of Nature*, p. 123), 'sound', and so on.
3. I am thinking in particular of Wahl's readers. Wahl's entire reading of *Process and Reality* turns around the question of the 'concrete events' in *The Concept of Nature*. But one could also include the readers of Cesselin who, in *La philosophie organique de Whitehead*, identifies 'actual entities' with 'events'.

4. Whitehead makes a distinction between 'physical prehensions', 'prehensions whose data involve actual entities' (PR, 23) and 'conceptual prehensions' which concern eternal objects.
5. See Christian, *An Interpretation of Whitehead's Metaphysics*, p. 202.
6. Leclerc, *Whitehead's Metaphysics*, p. 177.
7. Plato, 'Timaeus', pp. 1254–5.
8. Plato, 'Timaeus', p. 1255.
9. This is a common theme in Whitehead. The valorisation of mathematics permeated the initial moments of Greek thought, the principal consequence of this being an abusive generalisation of the *deductive* relation of ideas: 'Under the influence of mathematics, deduction has been foisted onto philosophy as its standard method, instead of taking its true place as an essential auxiliary mode of verification whereby to test the scope of generalities' (PR, 10). This idea as an invariant led to the deductive relation taking the place of all possible variations. In this way, the limited character of the deductive model was forgotten, a model which was able to be only an 'auxiliary mode' of verification, not of invention or of metaphysical construction.
10. Christian, *An Interpretation of Whitehead's Metaphysics*, p. 197.
11. Stengers, *Thinking with Whitehead*, p. 302.
12. See Pradeau (ed.), *Platon.*
13. Leclerc, *Whitehead's Metaphysics*, p. 94.
14. Wahl, *Vers le concret*, p. 135. (Translator's note: my own translation.)
15. Stengers, *Thinking with Whitehead*, p. 303. See Stengers, *Penser avec Whitehead*, p. 340. (Translator's note: the final sentence of this quotation is my own translation since it is omitted from the English version.)
16. Plato, 'Timaeus', p. 1255.
17. I am using 'survey' [*survol*] in the sense given to it by Ruyer, who develops the notion in relation to the idea of 'theme'. See Ruyer, *La genèse des forms vivantes*, p. 37.
18. Leibniz, *New Essays*, p. 74.
19. Whitehead, *Science and the Modern World*, p. 159.
20. Whitehead, *Science and the Modern World*, p. 160.
21. Whitehead, *Science and the Modern World*, p. 160.
22. Lautman, *Essai sur l'unité des mathématiques*, p. 204. Connecting this research to the Heideggerian difference between Being and beings, Lautman makes Being into the object of the dialectic. 'The principal moments in the unveiling of Being are as follows: first and foremost, it comes from *posing a question* on the subject of some-

thing. This does not necessarily mean that we perceive the thing to which the question was posed in its essence [. . .] It follows, then, and this is the crucial point, that this unveiling of the ontological truth of Being cannot take place without concrete aspects of *ontic existence* being materialised' (*Essai sur l'unité des mathématiques*; p. 206; translator's note: my own translation). It is not my intention, here, to question the relevance of this relation between Heidegger and Plato. What is interesting, however, is the relation between a question posed and the events of its solution.

23. See Deleuze, *Difference and Repetition*, pp. 177–84.

9

Temporal Dimensions of Actual Entities

Becomings and Instants

The individuation of actual entities has a beginning and an end. It passes from disjunctive diversity, where it has the character of an 'aim', a potentiality, through its being in action, to its satisfaction, at which point it is no longer susceptible to transformations or becomings. The best way of expressing this highly delimited nature of individuation is to use terms like 'blocks', 'blocks of becoming' or 'blocks of individuation'.

In *Some Problems of Philosophy*, William James, attempting to account for sensible experience, for perception in immediate experience, develops a notion of what he calls 'drops of experience':

> Either your experience is of no content, of no change, or it is of a perceptible amount of content or change. Your acquaintance with reality grows literally by buds or drops of perception. Intellectually and on reflection you can divide these into components, but as immediately given, they come totally or not at all.[1]

Divisions within a 'drop of experience' are always possible, an analysis into parts or elements can always be performed, and yet to do so would be to lose what is important, namely, that *these drops are indivisible totalities*. The divisions are ideal; they emerge out of acts of representation which translate what is given in totality into distinct elements. The most concrete experience is that which can be expressed by movements like *augmentation*, *intensification* and *amplification*, movements which reach a point of effective realisation, a point which, at the same time, marks the passage to a new 'drop of experience'.

Events form sequences or successive series of 'drops', each with their own individuality, their own manner of being, their own points of intensity. This does not mean, however, that there is no relation between events. One drop succeeds another, but is linked to it. The entire interest and complexity of James's approach lies in knowing how to establish distinctions, ways of closing up events, at the same time as finding ways of linking them together. In a prior work, *The Principles of Psychology*, already concerned with the problems addressed in *Some Problems of Philosophy*, James gives an example:

> Into the awareness of the thunder itself the awareness of the previous silence creeps and continues; for what we hear when the thunder crashes is not thunder pure, but thunder-breaking-upon-silence-and-contrasting-with-it.[2]

The thunder forms an event, but it has meaning only through the difference it produces with the event of the 'silence' preceding it and the new silence following it. The elements of the series communicate through these 'contrasts' and repetitions, which are simultaneously internal and external relations, or, more properly, which stand apart from the difference between interiority and exteriority.

Whitehead thinks speculatively what James explores as a mode of experience of consciousness. The notion of 'blocks of becoming' – Whitehead prefers to call them 'atoms of becoming' or 'acts of becoming' (PR, 68) – is an attempt to account, in terms close to those of James,[3] for the limited character of individuation. The scale of the problem, however, as well as the way it is set up are different. The fact that James poses the question on the basis of perspectival experience leads him to take an interest in forms of continuity between drops of experience. This is made clear by the 'metaphor' of the drop, as if the drop's intensity and individuality could spread into and join up with new elements. When Whitehead speaks of 'atoms', however, it is in a radical sense: *there are no exceptions to the limits of becoming*. Both thinkers, however, converge around this idea that the 'nature' of time, the time of experience for James or the time of existence for Whitehead, must be situated in these primordial elements called 'drops' or 'atoms'. *Time, then, is atomic by nature*. Since the question of time is posed on the basis of individuation and is identified with it, how could it be otherwise?

The first consequence is that the conception of time in *Process and Reality* resembles certain elements of the classical conception of time as much as it criticises them. Whitehead finds the 'classical' image of time at work implicitly in vastly different domains and authors. It is expressed in Newton's *Scholium*,[4] in the idea of a simple localisation of matter within the modern sciences, but also in the philosophies of Descartes, Locke and even Hume. Whitehead refrains from offering a systematic description of these ideas, preferring rather to provide brief citations and examples that demonstrate a common way of conceiving the 'nature' of time. His critiques are always local. We can, however, try to bring to light what this approach to individuation implicitly rejects – according to the way it describes time – by exploring the most systematic consequences of one of these propositions.

It can be found in Descartes' *Principles of Philosophy* when he writes that 'the nature of time is such that its parts are not mutually dependent'.[5] What does Descartes understand by parts? He clarifies this in an example he develops in the *Meditations*: 'a lifespan can be divided into countless parts, each completely independent of the others'.[6] *A life would be a kind of series*, a sequence of events in which each part gives way to one following it. This movement provides the 'part' with its qualities: a part is *an independent element inscribed within a succession*, a kind of link that never gets linked up. The nature of time is thought through this independence of successive parts, expressed by Descartes in various ways: as a 'moment', as a 'now', although his most appropriate term is 'instant'. The fundamental dimensions of time lie in the qualities of the instant: *the past is nothing but an instant that has passed and the future is an instant that is to come.*

The entire question revolves around knowing how Descartes, proceeding from the notion of a *discrete* nature of time, can account for continuity. If parts, or instants, are radically independent from each other, how can we speak of *a* life or *a* duration? The answer, here, is less interesting than the way Descartes sets up the problem. First of all, he recognises the question's importance, declaring, 'it does not follow from the fact that I existed a little while ago that I must exist now'.[7] An instant in a life implies no necessity concerning the existence that follows it, since such a necessity would be a relation. As such, 'continuity' must be conceived as produced by an 'external and transcendent cause'. Existential persistence cannot be conceived 'unless there is some cause which as it were creates me afresh at this moment – that is, which preserves me'.[8] The pure succession of the series of instants is transcended by an external cause, namely God, who sustains the connection. Descartes' terms, here, are crucial: *continuity is thought as preservation, and this preservation is produced by a repeated act of creation.*

> For it is quite clear to anyone who attentively considers the nature of time that the same power and action are needed to preserve anything at each individual moment of its duration as would be required to create that thing anew if it were not yet in existence. Hence the distinction between preservation and creation is only a conceptual one [. . .][9]

Creation and preservation are identical, differing only in relation to a particular way of thinking. The exact same notion of power is at work in the creation of a substance as it is in the maintaining of its existence, in the preserving of its continuity through a succession of temporal acts.

Whitehead's atomistic conception, then, agrees with Descartes on two essential points: first, both share a *discrete* approach. Time is made of successive parts, segments, or acts. For Descartes, these parts are instants, whereas for Whitehead they are 'becomings' or 'blocks'.[10] Secondly, Whitehead agrees with Descartes on the idea that *continuity is a conservation and that this conservation is a creation.* Descartes searches for a necessary condition of the duration of substances, and finds such a foundation in God: it is God who maintains the continuity of substances through an act of repeated creation. Whitehead, of course, unambiguously diverges from Descartes in explaining the relation between the three concepts: continuity, conservation and creation. He refuses to shift the problem on to God. He refuses to leave substances behind to go looking for reasons lying behind conservation or creation. He instead takes hold of 'becoming' and never lets go.

Returning to Descartes is ultimately incidental if we are to consider the break Whitehead performs: the break is effected through a reappropriation of the discontinuity of the parts of time and of conservation through creation. All these elements must now be redeployed on to a new plane that would require neither the exclusion of the past and the future nor the search for an external and transcendent cause to explain their influence in the present.

The Temporal Thickness of Actual Entities

How do the present, the past and the future interact in individuation? We should begin by exploring what these dimensions of time correspond to in speculative philosophy, which will allow us to trace their mode of interaction.

The past corresponds to 'disjunctive diversity'. The latter is composed of all the actual entities that have reached satisfaction, that are fully realised; all the acts that have taken place. It is a minimal definition, constructed by Whitehead as follows: *we will call 'past' everything that precedes a certain individuation taken as a point of reference.* The past, then, appears fully relative: it has meaning and existence only in relation to a process of individuation. Because an entity emerges and tends towards its satisfaction, disjunctive diversity can be called 'past'. Without this, it would merely be everything that has taken place, without distinction, without a temporal scale for its existences. Another, more essential way of putting this is that disjunctive diversity accomplishes a function: it provides 'data' to new individuations, from which they are formed. An entity that has attained satisfaction is thereby added to diversity and does not disappear. It acquires 'objective immortality' through the totality of prehensions that could be made of it. Remember, an entity comes into existence through prehension, through the capture of disjunctive diversity. It follows, then, that the past can be defined as the 'real potentiality', the matter, of individuation. The past is this multiplicity of 'objective immortalities', of actions *qua* potentialities, decisions that direct future individuations. This is why there is no exteriority between the past and the new entity. The latter, through positive or negative prehensions, includes in its totality everything that composes it. It can be said, then, that the past is what never stops being enriched by new individuations while at the same time being actualised as an object in the present.

This integrated, actualised past, however, can only exist in a manner specific to the entity, specific to the 'how'. An actual entity derives passively from the past; its appropriation is specific to what it is, and especially to what it *aims* to be. The past could never be integrated if the entity were not animated by *something that does not come from its past.* This 'something', irreducible to disjunctive diversity, is the entity's 'subjective aim', its final cause. It is not the state that it will reach, a state that could be conceived independently of the process, as if its end were already given before its realisation. It is the 'aim' immanent to each prehension. What is more, it is precisely this pure aim that conditions its 'future'. The subjective aim, then, is not a moment to come, a being that could occur, but rather what actively determines and directs individuation. The future is nothing but a tendency, a movement internal to being. Again, beyond actual entities there is nothing.

The past and the future, then, are relative terms. They have no existence that would let them be considered 'in themselves', as modes of existence in their own right. The entire past exists within an entity, and the future gains consistency only when involved in concrete operations of coming into being. *The past provides the object of individuation and the future the goal of individuation.* The former is the active power that encounters the passive power of the latter. This precise point of interaction at which the two powers encounter each other is the present of concrescence, prehension itself. The present gathers the entire past by putting it into communication with the actual by way of a finality that does not emerge from it. There is, then, in every concrescence, in every 'temporal atom', everything that has ever occurred, as well as the irreducible novelty of a perspective that integrates the past according to a particular mode.

> Thus an actual entity has a threefold character: (i) it has the character 'given' for it by the past; (ii) it has the subjective character aimed at in its process of concrescence; (iii) it has the superjective character, which is the pragmatic value of its specific satisfaction qualifying the transcendent creativity. (PR, 87)

The first two characteristics return to what we have already seen. The third, however, has a peculiar status. It is no longer a question of accounting for what happens inside a temporal atom and what constitutes it, but rather of accounting *for what it produces beyond its own existence.* This characteristic no longer concerns the actual entity's individuation but rather its consequences *for* other things. This is what Whitehead means when he speaks of 'pragmatic value': *what are the effects of this 'temporal atom'*, of this existence? Here we find the principle according to which every entity is a decision imposed on all new individuations. This decision is 'how the actual entity, having attained its individual satisfaction, thereby adds a determinate condition to the settlement for the future beyond itself' (PR, 150). This value is transcendent because it is located in the repetitions that will be made of the actual entity, not directly within in.

The Epochal Theory of Becoming

I have now described the basic elements of time, the 'parts' that form its nature. But what about the series? At the heart of this question, as it was already for Descartes, is the passage from discontinuity to continuity. Whitehead expresses it with a proposition: '[t]here is a becoming of continuity, but no continuity of becoming' (PR, 35). Continuity, here, is revealed through a contrast, the formulation of which might seem perplexing. But this opposition simply repeats elements that have already been brought to light, and announces the problem that must now be thought. To say there is no 'continuity of becoming' is another way of affirming the atomic nature of time, that these atoms are becomings held within strict limits and which, in this way, lack 'continuity'. For Whitehead the problem is found in the first part of the proposition: what is a becoming of continuity?

In his response he refers to Zeno's well-known paradoxes concerning movement, and to Aristotle's descriptions of them in the *Physics*.

> Zeno's arguments about motion, which cause so much trouble to those who try to answer them, are four in number. The first asserts the non-existence of motion on the ground that that which is in locomotion must arrive at the half-way stage before it arrives at the goal.[11]

The paradox lies in the fact that the moving body is unable to come to an end since it first must complete half of the total distance, then half of the half, and so on until infinity. In other words, it cannot be in movement. Interestingly, Whitehead and Bergson give opposite readings of this paradox. For Bergson, the paradox lies, ultimately, in the spatialisation of time inherent in representation. Departing from the fact that 'it is possible to distinguish points on the path of a moving body', it follows that 'we have the right to distinguish indivisible moments in the duration of its movement'.[12] *The intellect effects a translation*: the duration of the journey is transformed into a multiplicity of instants corresponding to the multiplicity of points on a line. Once this translation has been effected it becomes impossible to think movement. As such, the Bergsonian solution passes through a deepening of time as *duration* and *continuity*:

> in order to avoid such contradictions as those which Zeno pointed out [. . .] we should not have to get outside of time (we are already outside of it!), we should not have to free ourselves of change (we are already only too free of it!); on the contrary, what we should have to do is to grasp change and duration in their original mobility.[13]

Whitehead agrees with this diagnosis. A kind of spatialisation can indeed be seen in these paradoxes, which is why he writes that 'on the whole, the history of philosophy supports Bergson's charge that the human intellect "spatializes the universe"; that is to say, that it tends to ignore the fluency, and to analyse the world in terms of static categories' (PR, 209). Bergson, however, does not stop there; he goes further and, according to Whitehead, too far. He 'conceived this tendency as an inherent necessity of the intellect' (PR, 209). It is as if this spatialisation were inscribed in the nature of the intellect itself as one of its essential qualities. Whitehead distances himself explicitly, here: 'I do not believe this accusation; but I do hold that "spatialization" is the shortest route to a clear-cut philosophy expressed in reasonably familiar language' (PR, 209). This is a recurring theme in *Process and Reality*, a consequence of the ambition of speculative philosophy itself: *concepts, or the intellect, have no stable 'nature', but are part of an open adventure*[14] in which spatialisation is a possible, though not an exclusive, path.

The principal divergence between Whitehead and Bergson is to be found not in the diagnosis but in the problem's resolution. For Bergson the solution passes through a deepening of duration, a transformation of the relation to experience, a metaphysics that puts whoever commits to duration into a relation of genuine sympathy with it. If the paradox is possible it is because representation never stops introducing distance into our relation with a continuous, moving reality: it translates the latter into its own terms and so introduces false problems (problems that emerge out of representation's own operation and projections). In short, in the Bergsonian approach the possibility of leaving the paradox behind involves a transformation, a sort of philosophical asceticism in which concepts are led towards something else, a sympathy, and point towards a reality that overflows their habitual use.

This reality, different from the image that representation produces of it, is that of *a continuity of events, deeper than all apparent discontinuity*, of a duration the stabilities and pauses of which are only partial and secondary aspects explicable by the duration itself. Speculative philosophy, then, obviously has little in common with this 'continualist' vision of duration. To think an atomised time, as Whitehead does, it must be possible to at once leave Zeno's paradox behind while refusing to make continuity into the primary dimension of the real. The problem has to be reconfigured.

> The argument, so far as it is valid, elicits a contradiction from the two premises: (i) that in a becoming something (*res vera*) becomes, and (ii) that every act of becoming is divisible into earlier and later sections which are themselves acts of becoming. (PR, 68)

The first premise is key: it reminds us that all becoming is the individuation of *something*. To posit a becoming in itself, without content, or with a content different from its existence, has no meaning. 'Becoming' and 'individuation' are one and the same thing. The second premise highlights that every action, every becoming, is divisible. Actions are blocks that, although divisible, are nevertheless still totalities and, as such, indivisible. The paradox, then, comes from a translation, though not a translation that would force reality to take the form of the intellect, but rather a translation that substitutes *the possible for the real*. Because a totality *can* be divided we conclude that it is *in fact* actually divided, composed of distinct and divisible parts.

The movement forms a series or a sequence of actions. Whatever part we arbitrarily indicate 'presupposes an earlier creature which became after the beginning of the second and antecedently to the indicated creature' (PR, 68). Each action, then, inherits what precedes it and what it includes. Whitehead also speaks of a *transmission of feelings*: the antecedent transmits its own existence to that which follows it and the latter takes over this inheritance for those that come after. It is this series of legacies, of repetitions and transmissions that he calls a 'route'. If the movement reaches its end it is because the movement is transmitted from one moment to the next, because each part of the route is the repetition of what preceded it and the passing on to what follows it, without any possible reversal. In identifying becoming with the 'becoming of something' and in turning these actions of becoming into successive actions, *Whitehead is able to escape the paradox at the same time as he accepts its terms*. The answer's complexity lies in its refusal to reduce the notion of a series of transmissions to a 'unique series', that of a homogeneous time:

> There is a prevalent misconception that 'becoming' involves the notion of a unique seriality for its advance into novelty. This is the classic notion of 'time', which philosophy took over from common sense. Mankind made an unfortunate generalization from its experience of enduring objects. Recently physical science has abandoned this notion. Accordingly we should now purge cosmology of a point of view which it ought never to have adopted as an ultimate metaphysical principle. In these lectures the term 'creative advance' is not to be construed in the sense of a uniquely serial advance. (PR, 35)

The Distinction between Two Fluxes

Modern philosophy, for Whitehead, could indeed have put in place a theory of becoming close to that of *Process and Reality*. In its most important forms, however, it is criticised: 'the group of seventeenth- and eighteenth-century philosophers practically made a discovery, which, although it lies on the surface of their writings, they only half-realized' (PR, 210). What is this 'discovery' that lay on the surface? What is this implicit conception that spans the philosophy of the seventeenth and eighteenth centuries and that *Process and Reality* attempts to explain and to systematise? Certainly not what has been analysed above, the notion that consists in turning the instant into the paradigmatic dimension of time, to the extent that other dimensions become watered-down expressions of it. What Whitehead reads in certain tendencies of modern philosophy is a distinction that is never made fully explicit, despite the fact that it is mobilised in a series of problems connected to, among others, the relations between the subject and experience.

> This implicit notion of the two kinds of flux finds further unconscious illustration in Hume. It is all but explicit in Kant, though – as I think – misdescribed. Finally, it is lost in the evolutionary monism of Hegel and of his derivative schools. (PR, 210)

The terms, of course, vary from one philosopher to the next, from Hume to Kant, which is to say that outlining them requires very precise work, not only concerning vocabulary but also the tendencies of the problem. We can limit ourselves to Hume, seen by Whitehead as the precursor of his own distinction between the two types of flux, even if, following Hume, this distinction remains unconscious: 'From the point of view of the philosophy of organism, the credit must be given to Hume that he emphasized the "process" inherent in the fact of being a mind' (PR, 151). The first flux is that of the experience the subject has of his or her own perceptions in so far as they change from one to the next. Hume summarises this experience in a fundamental passage from the *Treatise of Human Nature* in so far as it concerns 'personal identity'. I reproduce it below in its entirety since it condenses the parts of the *Treatise* about the question of the 'self', articulating the experience of the flux of thought that matters to Whitehead.

> I may venture to affirm of the rest of mankind that they are nothing but a bundle or collection of different perceptions, which succeed each other with an inconceivable rapidity, and are in a perpetual flux and movement. Our eyes cannot turn in their sockets without varying our perceptions. Our thought is still more variable than our sight; and all our other senses and faculties contribute to this change; nor is there any single power of the soul, which remains unalterably the same, perhaps for one moment. The mind is a kind of theatre, where several perceptions successively make their appearance; pass, re-pass, glide away, and mingle in an infinite variety of postures and situations. There is properly no *simplicity* in it at one time, nor *identity* in different; whatever natural propension we may have to imagine that simplicity and identity.[15]

For Hume the experience of flux appears to be based on perceptions, impressions and ideas. It is relative to changing contents. One perception succeeds another in an infinite variation, one thought imposes on and substitutes itself for another through the combined activity of experience and the faculties. Flux, then, is not continuous and undifferentiated: it is a 'perpetual' movement of individual and distinct elements. We are far, then, from the idea of a 'continuum': this idea is precisely situated in the relations between fundamental elements of experience.[16] I have already shown how Hume describes *an essentially restrained kind of flux*: it is composed of 'parts', of 'elements', of 'pieces'. Whitehead calls it 'transition'. Hume says that this is the first form of flux, the passage (transition) of one element to another.

Hume, however, adds a second kind of flux to the first, and if he holds back from truly distinguishing it from the former, he nevertheless mobilises it in his examples and in his theory of the constitution of complex ideas, in which it is omnipresent. It concerns not only transition but *the elements of this transition*, everything described as the 'parts', 'elements' and 'pieces' that make transition possible. It is not only the passage from one part to another that matters, but the constitution of each part. The central term in Hume, then, is no longer the transition or the passage but the 'synthesis': passive syntheses of habit that spontaneously associate and link together the collection of impressions that furnish the senses, syntheses of the imagination that can form complex and unified ideas from simple ones.[17] There would be a kind of permanent pulsing of experience, moving from one synthesis to another. Still in the same part of the *Treatise* Hume writes that

> [o]ur impressions give rise to their correspondent ideas; and these ideas in their turn produce other impressions. One thought chases another, and draws after it a third, by which it is expell'd in its turn. In this respect, I cannot compare the soul more properly to any thing than to a republic or commonwealth, in which the several members are united by the reciprocal ties of government and subordination, and give rise to other persons, who propagate the same republic in the incessant changes of its parts.[18]

A multiplicity of syntheses of the soul are like distinct 'persons' who maintain relations with one another and can themselves be the object of new syntheses overflowing their own beings.

For Whitehead, then, Hume's distinction between two fluxes is expressed by two concepts: *transition* and *synthesis*. Synthesis is the constitution of a unity of experience, of a part of it; transition is the passage from one part to another, from a constituted synthesis to a new synthesis in the process of realisation. As always, Whitehead's reader encounters a significant dislocation of the Humean problem, a retranslation driven by the problem's potential importance for speculative philosophy. On this basis, on the basis of a problem that is not strictly Hume's, Whitehead is able to say that '[Hume's] analysis of that process is faulty in its details. It was bound to be so; because, with Locke, he misconceived his problem to be the analysis of mental operations' (PR, 151). If Hume is mistaken, then, it is not because he has made errors. It is because he remains wedded to the problem of classical empiricism: an analysis of the understanding. It has now, however, been amply reiterated that the speculative problem is constituted not by an analysis of perceptual experience, nor by a philosophical anthropology, nor by a relation to given experience, but rather by an approach to existence as such. Speculative philosophy, then, 'disagrees with Hume as to the proper characterization of the primary data' (PR, 151). These initial data are the impressions that Hume considers 'original' and from which he excludes all genesis. Whitehead, however, thinks becomings.

These two critiques can be brought together: it is because Hume locates the problem of flux in the understanding that he comes to see impressions as 'original'. If he had left the understanding behind, if he had posed the question on the plane of existence itself, he would have had to accept, according to Whitehead, that what is 'original' is simply the by-product of previous operations of synthesis. It could be said, then, that 'impressions' are the equivalent of 'satisfied' actual entities, and that the mind is the equivalent of the subject of a new synthesis.

> In the philosophy of organism, 'the soul' as it appears in Hume, and 'the mind' as it appears in Locke and Hume, are replaced by the phrases 'the actual entity', and 'the actual occasion', these phrases being synonymous. (PR, 141)

It is possible, then, to return to Hume's distinction by transforming its terms of classical empiricism into the language of speculative philosophy, in which what matters is existence as individuation.

> One kind is the concrescence which, in Locke's language, is 'the real internal constitution of a particular existent'. The other kind is the transition from particular existent to particular existent. This transition, again in Locke's language, is the 'perpetually perishing' [. . .] (PR, 210)

The relation between 'transition' and 'concrescence', between passage and unification, is a relation Whitehead calls rhythmic: '[t]he creative process is rhythmic: it swings from the publicity of many things to the individual privacy; and it swings back from the private individual to the publicity of the objectified individual' (PR, 151). The first is concrescence and the second is transition. Recall that the term 'rhythm' comes from the Greek *ruthmos*, which is composed of two parts: first, *ekhein* means at once 'to have', 'to hold', 'to possess' and 'to maintain'. These are the entities that become objects, that acquire objective immortality. But rhythm also refers to *rheo*, the flow and flux, the fluidity of becoming.[19] *A rhythm is a particular relation between discontinuity and continuity*, a passage between heterogeneous blocks. Wahl summarises the two fluxes and this construction of temporality as follows:

> In reality, temporalisation is made of discontinuous pieces of continuity. Temporalisation is atomic succession [. . .] As Zeno saw, there can be no continuity of becoming. What there can be is a becoming of continuity, a continuity formed, bit by bit, out of discontinuity.[20]

Notes

1. James, *Some Problems of Philosophy*, p. 80. Cited in PR, 68.
2. James, *The Principles of Psychology*, p. 240.
3. See Eisendrath, *The Unifying Moment*.
4. In that work Newton, writing on the subject of time and space, distinguishes between the absolute and the relative: 'I. Absolute, true, and mathematical time, of itself, and from its own nature, flows equably without regard to anything external, and by another name is called duration: relative, apparent, and common time, is some sensible and external (whether accurate or unequable) measure of duration [. . .] II. Absolute space, in its own nature, and without regard to anything external, remains always similar and immovable. Relative space is some movable dimension or measure of the absolute spaces; which our senses determine by its position to bodies, and which is vulgarly taken for immovable space' (Newton, *The Mathematical Principles of Natural Philosophy*, p. 77, cited in PR, 70).
5. Descartes, 'Principles', p. 200.
6. Descartes, 'Meditations', p. 33.
7. Descartes, 'Meditations', p. 33.
8. Descartes, 'Meditations', p. 33.
9. Descartes, 'Meditations', p. 33.
10. It is interesting to note that Whitehead's conception has not remained the same from one work to the next. In *The Concept of Nature*, for instance, he offers precisely the opposite idea: what comes first is

the continuity of what he calls the 'passage of nature'. This passage is a continuous movement that we experience in perception. What we distinguish, or differentiate, are 'events' which are like 'parts' or 'pieces'. The expression 'parts', here, might lead one to believe that there are neat distinctions between events, a border separating one from the other. But there is nothing of the sort: we pass from one event to the other by degrees, not by ruptures; the limits separating them are never cleanly cut since one event continues into another, it impinges upon the other's existence and persists beyond our actual experience of it. Whitehead speaks of their *extension*. In this way, the problem is posed in an opposite way to *Process and Reality*: continuity comes first in the form of a passage, and discontinuity is an effect of our experience of nature. Whitehead's reversal of perspective seems surprising. How can he take opposing points of view from one book to the next? It is enough, however, to recall that the object of *The Concept of Nature* is perceptual experience. It is important to remember this, since it gives an indication that Whitehead has changed planes, not conceptions of time – he has moved from the plane of perception to the plane of speculation. Schematically, then, one could say that Whitehead passes from the analysis of the time experienced in perception to the analysis of a speculative or metaphysical time.

11. Aristotle, 'Physics', p. 891.
12. Bergson, *Matter and Memory*, p. 192.
13. Bergson, *The Creative Mind*, p. 117.
14. 'I am also greatly indebted to Bergson, William James, and John Dewey. One of my preoccupations has been to rescue their type of thought from the charge of anti-intellectualism, which rightly or wrongly has been associated with it' (Whitehead, *Process and Reality*, p. xii).
15. Hume, *Treatise*, p. 165.
16. Brahami elaborates this emergence of the self very well. He writes: 'The mind is not, in fact, anterior to perceptions. Instead of the "I think" being necessary for the existence of perceptions, it is merely an effect implying them. The "I" itself is nothing but a fictional way of thinking a cluster or a collection of perceptions as a perfectly determined whole [. . .] Starting from perceptions, then, means choosing not to start from either the perceiving subject or the perceived object' (Brahami, *Introduction au Traité de la nature humaine de David Hume*, pp. 42–3; translator's note: my own translation).
17. See Hume, *Enquiry*, p. 37.

18. Hume, *Treatise*, p. 170.
19. See Sauvanet, *Le rythme grec d'Héraclite à Aristote*, pp. 8–9.
20. Wahl, *Vers le concret*, p. 148. (Translator's note: my own translation.)

Part III

Experiences and Societies: Thinking Nature

10

A Universe of Societies

Actual Entities and Societies

'Actual entities' are at the centre of speculative philosophy. They allow us to say, not what existence is as it is felt, experienced or lived, but what *existence is as such*, in its own reality, 'in itself'. Whitehead is a metaphysician, for whom the philosophical question that gives meaning to *Process and Reality* remains: what is existence? The answer to this question is given in the very first pages of *Process and Reality*, as if it required no a priori justification, no demonstrative development passing through a chain of reasons and principles. The point is not so much to offer an answer as it is to give meaning to that answer, to unfold its full significance. This answer, analysed in the previous section, takes the form of a definition: *existence is an actualisation of creativity*. I have approached this definition in its purest abstraction, in the constraints it places upon the scheme's fabrication, without attempting to make the definition correspond with some reality or with particular aspects of our experience. No example from or evocation of our experience could account for what an actual entity is. This is because the concept escapes all inscription within the apparently obvious or familiar.

For Whitehead – although he remains extremely allusive on this question, to the extent that one can only offer a hypothesis – everything happens as if existence, defined on the basis of actual entities, is precisely *that which we do not and cannot experience*. The confusion between the ultimate elements of existence and experience is 'the mistake that has thwarted European metaphysics from the time of the Greeks'.[1] This error, at the foundation of a series of false problems, principally consists of transposing what is given in experience on to the plane of ultimate existential principles. Relations of resemblance and analogy are privileged, leading to a situation in which the first principles of a metaphysics are, to greater and lesser degrees, simply unconscious generalisations of experience or, more precisely, visual experience. Whitehead attempts to bring the distinction between existence and experience to light in order to resist this projection, this translation of our modes of experience into a notion of existence not based on the model that is of importance here.

In the introduction to *The Creative Mind*, Bergson highlights a particular relativity of experience not unlike what Whitehead aims to express:

> the world in which we live, with the actions and reactions of its parts upon each other, is what it is by virtue of a certain choice in the scale of greatness, a choice which is itself determined by our power of acting. Nothing would prevent other worlds, corresponding to another choice, from existing with it, in the same place and the same time [. . .][2]

For Bergson there is a world of experience linked to a *scale*, to an order of magnitude, itself dependent on a particular 'power of acting'. This does not mean that, having become aware of the arbitrariness of the origin of a particular world of experience that seems familiar, we can change that world according to our will or pass from one world to another. A 'power of acting' gives access only to one or two of these worlds, or perhaps, with the power of imagination, to a few others at the most. It would be absurd to claim to be able to embrace all of these worlds. Such a claim, at best, would be an affirmation of a multitude of 'powers of acting', denying, in this way, the specificity of what we are and of what matters to us. At worst it would be to reduce all others to the familiar image of our own.

In the framework of speculative philosophy one might say that all worlds require the notion of the actual entity, that, despite their heterogeneity, they are all composed of *the same stuff*. It does not necessarily follow, however, that there is a hierarchy between them, that some would be better than others at expressing what actual entities are. *No world of experience can claim a direct resemblance with actual entities.* Actual entities are specific modes, particular organisations. Whitehead offers a technical term for expressing the elements of these worlds of experience that fail to correspond directly with actual entities. He calls them 'societies'. 'Society', here, should be taken in a broad sense, without any implicit reference to anything 'society' might mean in everyday language. Realities as diverse as an atom, a cell, an impression, an object, an individual and even a civilisation are called 'societies'. Everything that *has experience* and everything that *is experienced* can be called a society; the two are never clearly distinguished. This concept is at the centre of the highly distinctive form of empiricism at work in *Process and Reality*, since its aim is to construct a notion of experience to be deployed in speculative philosophy. There is, in the theory of societies, a crucial attempt to reconstruct such a concept. It is all the more surprising, then, that the concept of societies has played such a minor role in the majority of commentaries on *Process and Reality*.

Whitehead also speaks of the scale at which societies are found, a scale he terms 'macroscopic' to distinguish it from the scale of actual entities for which he reserved the term 'microscopic'. Do not be mistaken, however: the difference between the two is not a question of degree, *quantitative*, presupposing an order of magnitude going from the smallest to the biggest. It is *qualitative*: the microscopic is that which we can never experience directly, whereas the macroscopic is that which can be an object of experience (which is not to say that it necessarily is). There are distinct qualities separating the two orders, characteristics which, at certain points, are completely opposed. I am in full agreement with Leclerc when he explains the confusion – already raised above – between existence and experience by characterising it in terms of a relation between actual entities and the macroscopic.

> Whitehead points out that one prominent, if not predominant, tendency in the philosophical tradition has been to take the macroscopic objects of experience as instances of actual entities, and to generalize from these. It is to these entities [. . .] that the traditional Aristotelian logic is primarily applicable.[3]

What is the mode of existence of 'societies'? Where do they come from? What composes them? Whitehead's answer brings no new elements to speculative philosophy: a society is a cluster, a collection, or, more exactly, a togetherness of actual entities. A society is nothing but a *relation* that makes something that matters from certain aspects of the modes of existence of actual entities. Societies cannot be identified with actual entities since the latter are modes of existence that cannot be limited to societies, and, reciprocally, societies are not reducible to the qualities of actual entities. It is as if actual entities have particular ways of being that produce new scales of emergent qualities distinct from them. 'Societies', then, are the effects of actual entities. 'Effect', however, should be taken in an empiricist sense, as when Hume writes that the 'effect is a distinct event from its cause'.[4] It is the key question, at the foundation of *Process and Reality*'s empiricism: how do existences produce elusive forms of experience that fail to directly correspond with them?

Before attempting an answer, the existential specificity of 'societies' should be clarified. Whitehead writes that '[a]n ordinary physical object, which has temporal endurance, is a society' (PR, 35). It can be concluded from this that what brings together an idea of an object, an individual or a living being, however ephemeral it may be, is that, in some way, whatever its form or way of occupying space, it *endures*. A minimal definition of the concept of society, then, can be brought together with that of duration. In the same way that becoming and individuation can be identified with actual entities, duration and persistence should be identified with societies. This is a difference in scale, and should be taken as a radical refusal of any comparison or analogy: 'The real actual things that endure are all societies. They are not actual occasions.'[5] To account for experience as a multiplicity of societies, then, requires explaining the *constitution of durations* and their interconnections.

The Emergence of Societies

The basis for the emergence of societies should be located in the mode of functioning of actual entities. Remember that the first plane of existence, or as Whitehead calls it, the first dimension of being, is that of a pure disjunction of already existing actual entities: 'disjunctive diversity'. As already shown, disjunctive diversity forms the 'real potentiality' for individuation from which new entities are created via prehensions. It has nothing to do with any kind of merging, gathering or totality but rather with real disjunction, the being-disjointed of actual entities.

The first form of connection within diversity Whitehead calls a 'nexus'. This is the most minimal form of a togetherness of several entities. Remember that an actual entity already constitutes a togetherness. A nexus, however, is a kind of merging together of actual entities. A nexus involves only a highly limited section of 'disjunctive diversity', since not all entities are necessarily part of the nexus. Actual entities in disjunctive diversity can exist outside of any association, free from all relations with others. An actual entity may belong to several nexūs simultaneously, linked in diverse forms of affiliation to a multiplicity of gatherings of actual entities. None of these forms of existence of entities can be excluded a priori; they can be important whether or not they are independent of a nexus or a multiplicity. On the level of disjunctive diversity, the nexūs appear as transitory and local forces in all cases, points of merging in a continuous flux of geneses and becomings that overflows them on every side.

This is how Whitehead characterises a nexus, as a form of relation: it is 'a set of actual entities in the unity of the relatedness constituted by their prehensions of each other, or – what is the same thing conversely expressed – constituted by their objectifications in each other' (PR, 24). A nexus is the product of a *mutual operation* of prehension through which a common past is prehended and inherited. A nexus is formed as soon as a reciprocal prehension takes place within diversity, whatever the number of terms engaged in that relation. Further, the concept of a nexus involves no new concept; no new element supplements the processes of individuation. It aims only at expressing those particular cases in the order of existence where, first, *subjects simultaneously prehend the same objects*, secondly, *objective existences integrate the same formal existences*, and finally, *passive powers are determined by the same active powers*.

'There are thus real individual facts of the togetherness of actual entities, which are real, individual, and particular, in the same sense in which actual entities and the prehensions are real, individual, and particular' (PR, 20). In this sense, nexūs are real forms, just as real as actual entities themselves. The link that unites them in this form of existence is neither superficial, nor secondary, nor illusory: it is the source of a mode of being concerning disjunctive diversity. Whitehead also calls them 'non-social nexūs'. Of course, as a grouping of actual entities a nexus is 'social' by definition: it is a *togetherness*. The reason Whitehead emphasises the non-social operation in this form of association is that he reserves particular qualities for the term 'social' that are not found in nexūs. The term 'social' is given a technical meaning: it is not redundant in relation to the term 'nexus' even if both emphasise the association and merging of actual entities. For a nexus to be social, it must fulfil three precise conditions:

> A nexus enjoys 'social order' where (i) there is a common element of form illustrated in the definiteness of each of its included actual entities, and this common element of form arises in each member of the nexus by reason of the conditions imposed upon it by its prehensions of some other members of the nexus, and these prehensions impose that condition of reproduction by reason of their inclusion of positive feelings of that common form. (PR, 34)

Only certain nexūs satisfy these conditions. In fact, social nexūs represent a very limited group of nexūs, those which already constitute singular forms of existence within disjunctive diversity. Social nexūs are the first condition of experience, testifying, in this way, to their specificity in relation to existence.

1. A nexus is social when 'there is a common element of form' in each of the actual entities that compose it. What Whitehead calls an element of form is what, at the level of actual entities, falls under the *manner* or the *how*, that is, the mode of prehension. The salient feature, here, concerns the participation in the same form by the group of entities composing the nexus. They all prehend in a similar manner, a manner distributed in the group of the nexus and that defines each entity as being in a singular relation with the others. The fact, however, that this is a *common* element of form does not imply that the entities are similar: they have a common form, but the processes of individuation are all specific. To understand this relation between the 'common element' and the specificity of the individuation, remember that the manner in which an entity prehends depends upon the ingression of multiple eternal objects that determine the type of prehension that the actual entity performs upon the data preceding it. It could be said, then, that the element of common form to the actual entities of a social nexus is a complex eternal object, 'exemplified in each member of the nexus' (PR, 34).

2. The reason for the distribution of this 'common element of form' must be sought not in an imposition coming from outside the nexus, constraining it in a particular way, but rather in requirements that each member places on the others. Actual entities are active powers that direct and channel becoming. They determine

creativity and, as such, determine their own objectification within new actual entities of which they then become ingredients. It follows, then, that every actual entity, whether inside a nexus or not, depends on what exists.[6] But the particularity of a social nexus is that these constraints are co-determining, they apply to each other and produce reciprocal relations. The common form is necessary because it is transmitted to every level of the nexus's existence. Every member embodies a constraint and a power for the others.

3. This 'common element' is repeated, transmitted across a 'historic route' proper to the nexus. This should be understood as a constraint on reproduction: the nexus reproduces its common form at every stage, but since its actuality depends on what precedes it, it inherits the constraints of prior existences. This is the radical difference between social and non-social nexūs. Whereas the latter emerge as essentially momentary forms of relation between actual entities, the former repeat those relations, making them endure. A social nexus, then, is a system of transmissions of relations, inheritances and repetitions.

These are the three elements that define 'social nexūs' and that differentiate them from disjunctive diversity and non-social nexūs. It would be pointless to look for a foundation, a general a priori explanation for the emergence of social nexūs. Whitehead's empiricism emerges out of his initial refusal of all demarcations or explanations that would reduce the event-driven nature of emergence. All that can be said is that within disjunctive diversity there are convergences in certain areas, that within the multiplicity of prehensive operations characterising diversity there are reciprocal prehensions of a common past that impose a form and transmit that form to others. Nothing is required, here, beyond the modes of existence of actual entities in the diversity of their functioning. At issue was knowing what a 'society' is. A society is a 'social nexus': '[s]uch a nexus is called a "society", and the common form is the "defining characteristic" of the society' (PR, 34). These characteristics are generic; they concern all social forms at all levels of complexity, from an atom to the universe:

> Thus a society is more than a set of [actual] entities to which the same class-name applies: that is to say, it involves more than a merely mathematical conception of 'order'. To constitute a society, the class-name has got to apply to each member, by reason of genetic derivation

from other members of that same society. The members of the society are alike because, by reason of their common character, they impose on other members of the society the conditions which lead to that likeness.[7]

What is a Multiple Being?

The question, now, is to know where the identity of a society is *located*: at the level of actual entities or at the level of their relations? At the level of transmission or the constraints placed on that transmission? Within the 'common element' of form itself or in its distribution? In Locke's *An Essay Concerning Human Understanding* we find related, almost analogous, questions. This provides extra confirmation, if any were needed, of the importance of empiricism – and particularly Locke's version of it – in *Process and Reality* and its approach to societies. To set up the problem in such a way that would be appropriate to the question of societies, Whitehead, by way of fragments and allusions, repeats and transforms certain of Locke's passages. We should try, then, to re-establish in a more explicit way what Locke lays out. What about it inspired Whitehead to translate it into a more speculative language?

The key chapter of the *Essay* concerning societies is entitled 'Of Collective Ideas of Substances'. Here Locke distinguishes two ideas of substance: complex ideas of singular substances and collective ideas of substance. The first are those 'of several single Substances, as of Man, Horse, Gold, Violet, Apple, *etc*'.[8] Locke calls them complex ideas in so far as they incorporate a multiplicity of different ideas that affect the mind: colour, hardness, shape, sound, taste, and so on. Although they are comprised of multiple ideas, we can, each time, represent *a* 'man' or *a* 'horse' as unique entities as if they were simple and autonomous beings. But these beings are not *the one of the multiple*, as if the word 'man' were the *truth* of the multiplicity of ideas that it encompasses. They are, rather, existential unities *of the multiple as multiple*. Locke transforms classical perception, according to which the initial element is the individual to which a multiplicity of qualities is then attributed. The entire order of predication has been reversed: there is no longer a subject to which qualities are then predicated but rather qualities to which a subject is then attributed. This 'attribution' of an individual existence, the fact that we can say *this* man without necessarily examining the group of elements that compose the idea of him, is connected, for Locke, to the repetition and association of ideas.[9]

What separates Whitehead from Locke can already be seen in the terms of this distinction. For Locke, a multiplicity of initially simple ideas is constructed by the mind, which is to say the connection is external to the ideas.[10] In this way Locke mobilises *quantitative multiplicities*: groups composed of individual, independent and autonomous elements. For Whitehead, however, this nexus cannot be explained by any external function (such as the mind). It can be explained only by the immanent, prehensive activity of actual entities, as testified by the passages cited above. Every relation is immanent to its terms and each relation is internal to the society. These reservations, which seem like quibbles, crystallise the main differences between Whitehead and Locke. But they should not diminish what Whitehead does in fact inherit from Locke: the demand to think the multiple outside of an underlying unity, to think the fact, made evident by experience and observation, that there is nothing contradictory about saying *a* man, *a* rock or *a* particle at the same time as affirming that such unity is nothing but a multiplicity of elements connected in a certain way.

Locke, however, does not stop at the idea of multiplicity within individual substance. His thinking would, after all, have had little influence on *Process and Reality* if it were not connected to a more important idea. Alongside complex ideas of substance, Locke develops what he terms 'collective ideas of substance'. Whereas the former refer to a gathering together, a unification and composition of simple into complex ideas, the latter refer to the organisation of complex ideas themselves:

> the Mind hath also *complex collective Ideas* of Substances; which I so call, because such *Ideas* are made up of many particular Substances considered together, as united into one *Idea*, and which so joined, are looked on as one.[11]

Locke poses a genuine ontological problem, namely, that of complex organisations of beings, of the overlapping of scales. The idea of a man refers to a multiplicity of elements that the man holds together (complex ideas), but the man himself can be an element of another, much bigger grouping (collective ideas). Locke offers a series of examples that can all be read as examples of societies in the Whiteheadian sense. It is not just the similarity of their examples that is surprising, however, but the similarity of the very terms in which they are proposed:

> the idea of such a collection of Men as make an Army, though consisting of a great number of distinct Substances, is as much an *Idea*, as the *Idea* of a Man: And the great collective *Idea* of all Bodies whatsoever signified by the name World, is as much one *Idea*, as the *Idea* of any the least Particle of Matter in it.[12]

These are groups that include other groups. At every level and at every scale – from an atom up to the cosmos – they form singular and individual unities that function as unique beings. As a result, 'a Troop, an Army, a Swarm, a City, a Fleet' are as much individual beings as they are collective ones.[13]

What makes this idea of groups of substances[14] so interesting in the theory of ideas? The answer is that here, for the first time, Locke is able to think a complex idea that is not directly reducible to its simple elements, as was the case in his notion of ideas of substance. He can now think an idea of a group in which every element is itself another group, without losing its unity, its substantial being (*a* town, *a* swarm, *an* army, and so on). Locke's notion runs into difficulties, of course. But it nevertheless offers a totally new conception of substance as the nexus of a collective unity and a vision of the universe as a multiplicity of collective ideas of substances, complete with affiliations, participations and groupings. He comes close to a cosmological vision in which the universe is composed of zones of organisation and of connections between those organisations.

All the same, Locke ultimately diminishes the significance of collective ideas to maintain coherence with the rest of his philosophy: he reduces them to mere representations.

> And, in truth, if we consider all these collective *Ideas* alright, as *ARMY*, Constellation, Universe; as they are united into so many single *Ideas*, they are but the artificial Draughts of the Mind, bringing things very remote, and independent on one another [. . .] into one conception, and signified by one name.[15]

Ultimately, collective ideas are nothing but 'the artificial Draughts of the Mind'. In this way Locke recalls the limits of an empiricism that applies only to perceptual experience. He ends up by reducing the problem, by leading empiricism on the path of philosophical anthropology. Locke, nevertheless, cut free the fundamental question that empiricism has never stopped repeating and rephrasing, trying to save it from reduction.

James takes up precisely the same demand that a multiple being must be irreducible to its terms, that it should have a specific mode of existence, a 'togetherness' that is neither the simple juxtaposition of its elements nor a mere mode of perception. In these collective beings James sees a requirement that Whitehead takes literally: the immanent activity of their constitutive elements. 'A social organism of any sort whatever, large or small, is what it is because each member proceeds to his own duty with a trust that the other members will simultaneously do theirs.'[16] Everything that characterises a social organisation, '[a] government, an army, a commercial system, a ship, a college, an athletic team, all exist on this condition'.[17] Each element participates in this collective being. No element has any individual identity outside of the assemblage's collective being, as in the case of Locke's examples. This is precisely what Whitehead means by nexus: a unity of existence produced by a set of elements and their reciprocal activities that prolongs its form through the decision embodied in the becoming of each entity.

Immanent Social Identity

We can return, now, to the question of where a society's individuality is located. Whitehead offers a partial answer in a passage from *Adventures of Ideas*: '[t]he self-identity of a society is founded upon the self-identity of its defining characteristic, and upon the mutual immanence of its occasions'.[18] An important consequence follows from this: the principle of identity is immanent to the society. As such, and still in *Adventures of Ideas*, Whitehead writes: '[t]he point of a "society" as the term is here used, is that it is self-sustaining; in other words, that it is its own reason'.[19]

Properly speaking, only entities can be described as *causa sui*, self-caused, entities which, at the same time, are the causes of other dimensions of experience. When a society emerges, however, it exists, at its own scale, in a mode that might give the impression of being *causa sui* in the manner of Spinozist substance. Although societies require actual entities and other societies, they seem to lead their own lives, they appear 'self-sustaining', at least with respect to their own identities. Their unities are produced through the self-consistency and self-production of actual entities. They would be incompatible, then, with any explanation based on the nature of an essence that would exist prior to its composition.

To understand the identity of societies it should be asked not what they are but what they consist of. Identity must be located exclusively in the society's 'consistency'. In its primary meaning, to 'consist' refers to a notion of 'being composed of', 'including in one's composition', that is, to a particular relation of *affiliation* and *possession*. Every society, then, may be inventoried. We have already raised *Process and Reality*'s substitution, in its treatment of actual entities, of a logic of having for the more traditional logic of being. The theory of societies intensifies the supremacy of having over being by placing identity itself within a logic of having. One might say, then, that to consist of is to have, though in the sense in which what is possessed defines the society itself. Secondly, to 'consist' refers to *consistere*, meaning to 'hold together', to persist.[20] Bringing these two meanings together implies that the conditions of a society's definition is located in the form of a question that must be posed individually for each society: what does a society possess as its components and how do they hold together?

We can call this immanent connection (reciprocal prehension) the *self* of a society. Whitehead hesitates to call it a 'person'. As soon as there is consistency, there is something approaching what could be called a person. And yet 'unfortunately "person" suggests the notion of consciousness, so that its use would lead to misunderstanding' (PR, 35). This misunderstanding consists in describing every society as either *having consciousness* – or in the model of a consciousness that would be realised to a greater and lesser extent, according to the type of society – or as *tending towards consciousness*. In short, it would involve making consciousness into a general model, present at different levels in different societies. In the theory of societies, however, consciousness is one society among others. To be able to say 'I', or simply to be able to feel a distinction between what one is and what one experiences, requires initial activities that are themselves implicitly mobilised societies from which consciousness derives. Rather than a model, consciousness is simply a particular case. Societies do not aim towards consciousness, nor do they require any implicit consciousness. In certain more complex 'living' societies it could be said that, at the most, they express a 'self-enjoyment', a self-relation prior to consciousness and on which they are based. Remember what Deleuze writes of 'self-enjoyment', which is now particularly relevant: 'The plant sings of the glory of God, and while being filled all the more with itself it contemplates and intensely contracts the elements whence it proceeds. It feels in this prehension the self-enjoyment of its own becoming.'[21]

To avoid this misunderstanding Whitehead constructs another concept, closer to the immanent identities of societies and connected to their activity: in place of 'person' he suggests 'character': 'The nexus "sustains a character", and this is one of the meanings of the Latin word persona' (PR, 35). A character is defined by potentially variable roles and operations. This distinguishes it from the notion of a person which implies a permanence or an invariance underlying its movements. The 'character' is never given once and for all, not before nor following the actions that it performs, but *simultaneously*. It is not reducible to the logic of the a priori and the a posteriori but, it could be said, opens out on to a field of the *a praesenti* that captures some of its actions,[22] gradually constituting itself through choices and decisions. The most important distinction between the notions of person and character, then, is that the person is given directly, whereas the character *is constructed*, *acquires consistency*, a density of being intensified by its own operations.

This contrast between person and character can help us understand Whitehead's critique of the Cartesian cogito. According to Whitehead, Descartes conceives of the thinker as the origin of thought, as the person who thinks. For Whitehead, however, thoughts, in all their mutual operations, connections, reciprocal determinations, immanent constraints and the decision that each is for the other, produce the thinker. *A character is distributed in thought*. A character is constructed prior to the thinker by the immanent processes of thought inherited by that character. The cogito is a constructed society, composed of a multiplicity of other societies, of particular thoughts, memories or perceptions: these are all societies that define what the cogito is. Each of these societies produces a particular character, a consistency. But each one, at its own scale, forms a constraint upon and a possibility for the others, to the extent that the cogito should be seen as an assemblage of societies, a multiplicity of characters forming *unities of consistency*. And these unities, as for every individual society, are not the foundation of the cogito, nor are they the synthesis of all societies. They are fleeting states of balance.

James, in a famous passage in *The Principles of Psychology*, opposes a vision of a 'stream of thought' to this vision of the cogito. He develops an idea of impersonal thinking very close to that of the 'functioning' of societies.

> If we could say in English 'it thinks', as we say 'it rains' or 'it blows', we should be stating the fact most simply and with the minimum of assumptions. As we cannot, we must simply say that *thought goes on*.[23]

The cogito's reliance on the 'it thinks' should be thought, in its impersonal activity, as comparable to other events. The cogito entails only one additional thing: a return to the operations, an evaluation or judgement of them, a recovery or *repetition* of the activity. And yet the very state of being conscious of the activity of thought kicks off other operations, each with its own degree of impersonality. To ask oneself *what is thought?* is already an operation that involves the same principles as any other activity. It would be absurd, then, to try to reach *a degree of correspondence between the operations themselves and a consciousness of them*, to search for a maximal harmony between the two. In the same way that no entity can express its realisation without a new concrescence, a consciousness of the operations of thought can give rise only to new operations.

Societies like a 'rock', a 'molecule' and a 'man' are societies that contain a multiplicity of underlying characters in negotiation with each other. These societies are always 'common forms' that aim to persist for their own purposes. They are styles of connecting and reproducing themselves. Decisions that correspond to each society determine the whole of that society, through the expression or refusal of possibilities (positive and negative prehensions).

Notes

1. Whitehead, *Adventures of Ideas*, p. 204.
2. Bergson, *The Creative Mind*, p. 45.
3. Leclerc, *Whitehead's Metaphysics*, p. 119.
4. Hume, *Enquiry*, p. 27.
5. Whitehead, *Adventures of Ideas*, p. 204.
6. See PR, 83.

7. Whitehead, *Adventures of Ideas*, pp. 203–4.
8. Locke, *Essay*, p. 317.
9. See Locke, *Essay*, pp. 398–401.
10. See Michaud, *Locke*, pp. 136–45.
11. Locke, *Essay*, p. 317.
12. Locke, *Essay*, p. 317.
13. Locke, *Essay*, p. 318.
14. Leibniz speaks of 'the unity of the idea of an aggregate'. He reduces them to mental unities: 'So the only perfect unity that these "entities by aggregation" have is a mental one, and consequently their very being is also in a way mental, or phenomenal, like that of a rainbow' (Leibniz, *New Essays*, p. 146).
15. Locke, *Essay*, p. 318.
16. James, *The Will to Believe*, p. 29.
17. James, *The Will to Believe*, p. 29.
18. Whitehead, *Adventures of Ideas*, p. 204.
19. Whitehead, *Adventures of Ideas*, p. 203.
20. This is the definition of consistency appealed to by Deleuze and Guattari in *A Thousand Plateaus*: 'the "holding together" of heterogeneous elements' (*A Thousand Plateaus*, p. 323). They note the conditions of a philosophy of assemblages in which problems are no longer distributed on the a priori basis of a proper form, an essence or a being deeper than the grouping of connected elements. This is why an assemblage is defined only through the manner in which heterogeneous elements interact, and the more they are different the better they interact. An assemblage, for Deleuze and Guattari, is the minimal unity of existence at any register: 'The minimum real unit is not the word, the idea, the concept or the signifier, but the *assemblage*' (Deleuze and Parnet, *Dialogues*, p. 51).
21. Deleuze, *The Fold*, p. 78.
22. Here I take up Simondon's idea of the *a praesenti*. Simondon develops it to account for the relations produced within a process of individuation, relations that can be neither exclusively anterior, nor posterior, but simultaneous. As such, the process of individuation is neither 'a priori nor a posteriori but *a praesenti:* an informative and interactive communication between that which is bigger than the individual and that which is smaller than it' (Simondon, *L'individuation psychique et collective*, p. 66; translator's note: my own translation).
23. James, *Principles of Psychology,* vol. 1, pp. 219–20, cited in Lapoujade, *William James*, p. 29.

11

The Mode of Existence of Societies

Societies as Durations

The principal difference between 'actual entities' and 'societies' is that, while the former can only become, the latter persist, reach a stability of being. The former are 'blocks of becoming' or individuations, while the latter are 'durations'. For Whitehead, the great question is again expressed by Locke:

> But what Locke is explicitly concerned with is the notion of the self-identity of the one enduring physical body which lasts for years, or for seconds, or for ages. He is considering the current philosophical notion of an individualized particular substance (in the Aristotelian sense) which undergoes adventures of change, retaining its substantial form amid transition of accidents. (PR, 55)

Whatever the scale of this duration – years, seconds or centuries – the question is to know how a 'particular substance' *endures*, that is, maintains a form of identity across the changes affecting it. The manner in which Locke poses the question of duration and how he attempts to define it is crucial here. He doesn't say that duration could be measured on a uniform temporal scale, against a single measure that would apply to every substance. A thing that endures is not just something that occupies a longer or shorter measure of time. Locke's definition is entirely alien to more classical approaches to time: time, for Locke, becomes a *relation between persistence and change*. A substance is said to endure as soon as it changes without disappearing. The question has little to do, now, with a time that would simply be occupied by substances. The question, rather, is located within the substances themselves and their sensitivity – or lack thereof – to change.

If this conception of time is taken up at the level of societies – remember that Whitehead defines the latter as durations – then it might be said that a society is what *has the power to receive and to effect changes within itself*.[1] There is no hierarchy among societies. Although they can experience changes that last from a second to years and centuries, they all display the same characteristics. The question, however, is to know where to locate the powers of change and persistence of societies.

Whitehead refuses three solutions that emerge whenever the problem is posed: first, the answer that would diminish the question of transformation in favour of essence. If something were capable of transforming itself while maintaining an identity throughout that change then all changing qualities would be superficial and secondary in relation to a deeper identity, its essence.[2] Secondly: together with Locke himself, one might see in persistence a repetition of qualities to which the mind attributes an identity.[3] The reason for the persistence of a substance would not be immanent to that substance but must be found in the operations of a mind which, through imagination and memory, produces and maintains the identity of that substance. Thirdly: one could seek the foundations of persistence in another kind of reality. This, as shown above, was Descartes' solution, a solution that brought together conservation, creation and transcendence: substance is maintained and conserved by a repeated act of creation, performed at each instant, despite the changes affecting it throughout its history. For Whitehead, however, societies are essentially 'personal orders': 'an "enduring object", or "enduring creature", is a society whose social order has taken the special form of "personal order"' (PR, 34). The question of identity, then, of change and of persistence has to be located within this other, more technical question: what is a 'personal order'? Two conditions have to be met to speak of a personal order: 'when (α) it is a "society", and (β) when the genetic relatedness of its members orders these members "serially"' (PR, 34). Both are *relational* conditions. First, a society is, as shown above, a type of nexus, a form of relating, but with the specific requirement that its prehensions be reciprocal, that they descend from a common past. Secondly, Whitehead speaks of a *serial ordering*. This is also a relation, though of an entirely different kind: not *reciprocal* but *successive* prehensions. Each prehension succeeds others that have left a part of their being to it, the group coming to comprise a genuine 'historic route' that 'tends to prolong itself, by reason of the weight of uniform inheritance derivable from its members' (PR, 56).

The novelty of these constraints can be found, then, in this question of 'series'. It should not be confused with an order, a sequence of distinct elements connected by external relations, resemblances or rules. A series, for Whitehead, is not a particular determined relation between things. He immediately clarifies the two qualities by which the concept of a series is to be understood: *connection* and *genetic process*. A personal order is a series because the elements that compose it are involved in genetic relations and relations of connection with each other. To understand how it is possible to connect these two qualities and to use them to build the notion of a 'series', an example can be given that has had considerable sway in classical empiricism and that Whitehead takes up virtually unchanged. In *An Enquiry Concerning Human Understanding*, in the third section entitled 'On the Association of Ideas', Hume writes:

> Not only in any limited portion of life, a man's actions have a dependence on each other, but also during the whole period of his duration, from the cradle to the grave; nor is it possible to strike off one link, however minute, in this regular chain, without affecting the whole series of events, which follow.[4]

A man's life would be a series composed of links, an uninterrupted chain of distinct events. It would be located not in a certain event but in the movement that links events together, placing each element into a relation of dependency with all others. Hume's implicit definition of the notion of 'series', then, is found in the concepts of *succession* and *dependency*. In this sense, it might be said that Hume prefigures Whitehead's notion of the series as a historic route, as connection and genetic process. Looking closer, however, it quickly becomes clear that Hume's description of this dependency is, in fact, nothing of the sort. Hume does speak of a 'connection' between events, but this connection is itself derived; it rests on something deeper and more constitutive and to which, logically, it must be brought back.

What comes first for Hume is the *constant* and *repeated* conjunction of distinct and independent events. The idea that we can make a *connection* between things is an idea inferred from the repetition of a constant conjunction. In this way '[o]ne event follows another; but we never can observe any tye between them. They seem *conjoined*, but never *connected*.'[5] It therefore follows that 'we have no idea of connexion or power at all, and that these words are absolutely without any meaning, when employed either in philosophical reasonings, or common life'.[6] The example, then, can be reread on the basis of this clarification. The idea arises that a life is composed only of independent events to which we then attribute relations, connections and dependencies. The question of the 'series', of particular interest here, appears in Hume as merely a *succession* and a *conjunction*. This is a logically necessary consequence for Hume. Instead of an accidental fault in the system, it is instead the expression of its coherence. If Hume dedicates a section elsewhere to the question of 'necessary connection', placing it at the end of the sections that have laid out the principal workings of the human understanding, it is precisely because it appears to him as a major issue, a fundamental test of his theory of association and external relations.[7]

Whitehead takes up Hume's example for himself, mainly by substituting for Hume's conjunctions a genuine thinking of connection, bringing to light all the areas of dependency that, though accepted by Hume, ended up being diminished in favour of the distinctions between events. This is the passage in which Whitehead develops this relation between succession and connection most extensively. It contains one of the clearest descriptions of societies as series:

> For example, the life of man is a historic route of actual occasions which in a marked degree [. . .] inherit from each other. That set of occasions, dating from his first acquirement of the Greek language and including all those occasions up to his loss of any adequate knowledge of that language, constitutes a society in reference to knowledge of the Greek language. Such knowledge is a common characteristic inherited from occasion to occasion along the historic route. This example has purposely been chosen for its reference to a somewhat trivial element of order, viz. knowledge of the Greek language; a more important character of order would have been that complex character in virtue of which a man is considered to be the same enduring person from birth to death. Also in this instance the members of the society are arranged in a serial order by their genetic relations. Such a society is said to possess 'personal order'. (PR, 89–90)

A man's life is a series, itself composed of other series such as the knowledge of Greek, memories, things learned and impressions, but also biological and physical organisations which, in turn, can themselves be subdivided into a multiplicity of other series. I will return at greater length in this chapter to the relations between these series, their interpenetrations and dependencies. What matters for the moment, however, is this definition of a series as a *succession of genetic relations* in which each term is connected to and determined by all those that precede it. '[T]hus the defining characteristic is inherited throughout the nexus, each member deriving it from those other members of the nexus which are antecedent to its own concrescence' (PR, 34). It is precisely this relation of inheritance and tendency that defines the series, not any order of succession between otherwise unrelated elements.

The concept of 'genetic connection' aims simply to express Hume's example in its maximum metaphysical coherence: if a chain-link were subtracted, an event such as a life would be deeply altered. A society is indeed composed of actual entities and '[n]o actual entity can rise beyond what the actual world as a datum from its standpoint – its actual world – allows it to be' (PR, 83). It is inscribed within a history without which it would not exist and that each of its components embody. It is a complex system of relays: what is received is transmitted; the inheritance is included and then imposed on another entity. Every event is a succession of inheritances and constraints.

We can now return to the initial question. Duration is indeed a particular relation of change and persistence: change since the society never stops being modified by its own components (in particular by actual entities), and persistence through the inheritance of a common element along a historic route. This relation between persistence and change can be expressed by two classical terms, loaded with a history that can be recast in a single mode: reality and appearance. At the time Whitehead wrote *Process and Reality* the conflict between the pragmatists and certain forms of absolute idealism, such as Bradley's,[8] to which Whitehead explicitly refers, turned particularly around the question of the relation between 'appearance' and 'reality'. It could be said, in an obviously overly schematic way, that this opposition is between, on the one hand, a radical phenomenalism in which a logic of appearance or pure phenomenon is a genuine possibility, a phenomenon unrooted in any supposed reality outside of experience; and on the other, an absolute idealism in which appearance is relative to a finitude of experience, to our manner of being in relation to a real that overflows and grounds appearance.

Instead of two distinct registers of experience, in *Process and Reality* appearance and reality become two technical terms for expressing what belongs to a society. They become purely relative to the society in question. 'Reality' is the common and inherited form across the society's historic route, the movement of repetition and transmission established at the microscopic level of actual entities expressed by the society. Take Locke's example: the army that exists through the decisions of each of its members, that exists through what it maintains as an identifiable reality, *this* army that exists and shows itself. It might be arranged on a battlefield, it might be moving around or being transformed, ever changing, but it remains *this* army experienced by those who compose, encounter and perceive it. The same army can be experienced in the memories of its witnesses or historians. The army is a 'reality' in so far as it is a *unified* existence that endures along its own route. 'Reality', then, is not a category that comes under any specific mode of being: a memory, a book, a legend and an action are all 'realities' as soon as they emphasise their inheritance across a duration.

As for 'appearance': it in no way designates illusion, false or superficial perception – in speculative philosophy neither reality nor appearance are connected to perception – but rather the set of the society's variations, the decisions and immanent variations that do not affect its historical course. Extending the example of the army: it is continually affected by variations caused by the taking of leave, changes in troop-size, shifts in 'morale', changes of equipment and so on. These elements are 'appearances' across the army's historic route, since they concern the 'same' army, as long as these variations do not radically challenge its existence (through demobilisation, defeat, merging with a different army, and so on). If the character is the *reality* of the event, then its appearances are 'the differences between actual occasions in one event' (PR, 80). Whitehead writes in *Adventures of Ideas* that

> [a] society has an essential character, whereby it is the society that it is, and it has also accidental qualities which vary as circumstances alter. Thus a society, as a complete existence and as retaining the same metaphysical status, enjoys a history expressing its changing reactions to changing circumstances.[9]

This, in a highly distinctive mode, is a classical definition of substance as a non-changing reality with its own identity animated by 'accidental' qualities and variations connected across its history. Whitehead summarises this economy of 'appearance' and 'reality' as a relation between change and duration:

> Our lives are dominated by enduring things, each experienced as a unity of many occasions bound together by the force of inheritance. Each such individual endurance collects into its unity the shifting qualities of its many occasions.[10]

The Extensive Nature of Societies

Here, then, is a generic definition of the concept of 'society': a social order is repeated across a historic route. In this phenomenon there is no explanation that could attempt to found societies on, or distinguish them from, other forms of existence. It is an emergence connected to operations and encounters. Entities reciprocally prehend in certain situations, repeating prehensions of a 'common character'. This is an observed fact, and it should be given philosophical expression: in the universe there are events that emerge locally and provisionally through double captures [*entre-captures*] operated by actual entities. Speculative philosophy, here, joins forces with a radical form of empiricism. Taking this definition at its purest and most generic, however, only a small fraction of the forms that societies might take can be considered: those composed of actual entities. Such a mode of existence is fictional, of course: a society composed exclusively out of actual entities is never found in experience. This rough outline is necessary, though, as a basis for the construction of more complex societies that *do* determine our immediate experience.

The examples that Whitehead takes up, such as a man's life, implicitly express the 'content' of societies: a man is composed of organs, of cells, of molecules, but also of knowledge, desires, impressions, perceptions. If a man is a society, an event sustained across a historic route, then it follows that he is composed of a multiplicity of other societies,[11] themselves composed of other societies. A society, then, is an interlocking of sub-societies with tensions varying according to their individual natures. Remember that the term society is used only to make the multiplicity of modes of experience felt, modes that can only be pragmatically assessed. Certain societies, then, are composed of more or less autonomous sub-societies, while others exist only through powerful and interdependent tensions that they carry inside themselves. All societies, however, endure through inheritance, through the repetition of all of its members: entities and societies.

To understand this logic, we can detour to an author who, although he did not explicitly influence Whitehead – in fact, Whitehead seems to have ignored him – develops ideas that are surprisingly similar to this notion of societies. In his major work, *Life and Habit*, Samuel Butler writes:

> each cell in the human body is a person with an intelligent soul, of a low class, perhaps, but still differing from our own more complex soul in degree, and not in kind; and, like ourselves, being born, living, and dying.[12]

Cells compose the organs that house them, though cells are themselves composed of an ensemble of beings, particularly molecules.[13] Whitehead expresses this in a similar way when he writes that 'we speak of a molecule within a living cell, because its general molecular features are independent of the environment of the cell. Thus a molecule is a subordinate society in the structured society which we call the "living cell"' (PR, 99). Each one takes part in the existence of a society bigger than the organ that requires it. At the same time, it ignores the bigger society and operates for its own sake according to the logic and identity of its heritage and historic route. And if this can be said of molecules, 'so in all probability [could it be said of] separate electrons and protons' (PR, 99). We should be wary, however, of spatial expressions like 'in', 'within', 'inside' and so on; extensive relations are far more complex. These spatial forms are useful in examples such as the body and its organs, but they are unfit for other kinds of societies. A society that composes another is not necessarily inside the latter; it might even be excluded from it. Returning to the example of the army, it could be said that, as a society, it comprises a multiplicity of others (objects, organisational levels, reservists, and so on) that are not literally present in any demarcated geographical space.

Let us return to Butler. Following the passage commented on above, he clarifies the mode of existence of these 'persons':

> These component souls are of many and very different natures, living in territories which are to them vast continents, and river, and seas,

> but which are yet only the bodies of our other component souls; coral reefs and sponge-beds within us.[14]

What changes, then, is the entire notion of the individual. No longer simply that which moves through time, a primary identity with superficial variations and secondary changes, it is rather a set of *transactions*, of *negotiations*, of mutual demands and *produced dependencies* between existences, each one extending a history and inherited habits. Again, according to Butler:

> each individual may be manifold in the sense of being compounded of a vast number of subordinate individualities which have their separate lives within him, and with their hopes, and fears, and intrigues, being born and dying within us, many generations, of them during our single lifetime.[15]

Here is a better expression of what Whitehead understands by 'society': autonomous existences, marked by their history and animated by other existences that, most of the time, are ignored, though they would not exist without them. The difficulty lies in thinking these existences in relation – these 'souls', in Butler's terms – without building hierarchies or reductions.

Whitehead uses a term to summarise these relations of belonging among societies: extension. Extension is expressed through two fundamental relations: 'being composed of' and 'being a part of'. In this way every society has extension, it composes and comprises others. At the same time, however, it is animated by 'underlying' existences and is included inside larger ones. Stengers summarises these relations in the terms of a philosophy of events which, as seen above, is another way of expressing the philosophy of societies:

> the term 'extension' will be a primary term for characterizing the event *qua* connected. A discerned event always has an extension because it includes or comprises others, and it testifies to the extension of other events that include or comprise it. This is part of its meaning, as we are aware of it in perception.[16]

A 'discerned' event is an individual reality. It is always *this* rock or *this* man. Both, however, are composed of other events and form part of more extensive ones. Of course, the whole question, then, is knowing what relations exist between these societies, what relations there are between 'containing and 'being contained', between 'having' and 'belonging'. This distinction is always pragmatic, local and relational. A society that 'extends into others' is, in relation to those 'others', a 'structured' society in that particular and more or less provisional relation. Reciprocally, a society that composes another is, in relation to the latter, a 'subordinate' society. There is no 'ontological' difference here, no difference in reality between the societies, simply the bringing to light of extensive relations within certain existential situations. Whitehead summarises this economy of societies:

> A structured society as a whole provides a favourable environment for the subordinate societies which it harbours within itself. Also the whole society must be set in a wider environment permissive of its continuance. Some of the component groups of occasions in a structured society can be termed 'subordinate societies' [. . .] A 'structured society' may be more or less 'complex' in respect to the multiplicity of its associated sub-societies and sub-nexus and to the intricacy of their structural pattern. (PR, 99–100)

This distinction between 'subordinate' and 'structured' societies, however, might lead to a confusion that can be brought out in the form of a question: are structured societies the foundation, the explanatory principle of subordinate societies, or vice versa? This question reiterates, implicitly, the classical distinction of whole/parts. A structured society would be a kind of totality of existence formed by the 'parts' of subordinate societies. The answers to this problem are less interesting than what they presuppose: relations of 'belonging' would be explanatory relations, giving meaning either to parts or to wholes. If Whitehead had opted for one of the two solutions, however, he would have contradicted the principle of the identity of societies: *they are sufficient to themselves*, their identity is immanent to their functioning. A society, whatever its level of complexity, whatever the number of subordinates involved with it, is always its own reason for existing. It cannot be reduced to, nor based on, what composes it or what it composes.[17] The truth of the individual is not in nature, the truth of nature is not in the individual. The atom is not given meaning by the cell, the cell is not given meaning by the atom. When we speak of levels of being it is as genuinely non-reducible modes of existence; each level is a society and each society has an immanent principle of existence, an identity that crosses and defines it. And this principle is not to be confused with either a simple collection of constitutive elements, a pre-existing totality or a totality that it would stretch towards.

This does not mean, however, that societies never exchange with or determine others. On the contrary: to think in this way is to think connection. But this connection is always negotiated in some way: '[t]he doctrine that every society requires a wider social environment leads to the distinction that a society may be more or less "stabilized" in reference to certain sorts of changes in that environment' (PR, 100). The environment is the structured society to which another society is subordinated. Its relation to the environment is a relation established at once by the society's specificity and by the type of environment with which it interacts. This is why we speak of a link being *negotiated*: the link is not imposed on subordinate societies by structured societies but is characterised each time by the societies concerned.

It follows, then, that every event has a *significance* that we translate into our perspectival experiences as spatial depth. Events 'spread into other events, and other events spread into them'.[18]

> It should be possible to say that the event which was the assassination of Caesar occupies space. The relations of events to space and to time are analogous in almost every respect. There are not objects in space on the one hand and facts in time on the other. There are fact-objects, which are events.[19]

To compose or to possess are ontological characteristics as much as they are spatial ones. If an event can be defined by what it 'possesses', by what it includes as sub-societies, then it can also be defined by the space it unfolds. Both ultimately refer to the same idea: there is space only in the extension of an event. Or, more precisely, what we call 'space' is the expression of one of the possible forms taken on by the significance and extension of an event.

Notes

1. This is Aristotle's definition of passive powers. In book nine of the *Metaphysics*, Aristotle defines passive power as that which 'makes [a thing] capable of being changed and acted on by another thing or by

itself regarded as other' (Aristotle, 'Metaphysics', p. 3551). This is the principal object of Heidegger's course on Aristotle's metaphysics: Heidegger, *Aristotle's Metaphysics Θ 1–3*.

2. 'It seems most distinctive of substance that what is numerically one and the same is able to receive contraries' (Aristotle, 'Categories', p. 34).
3. In chapter fourteen of the *Essay*, a chapter dedicated to the question of duration, Locke writes: 'That we have our notion of *Succession and Duration* from this Original, *viz.* from Reflection on the train of *Ideas*, which we find to appear one after another in our own Minds, seems plain to me, in that we have no perception of *Duration*, but by considering the train of *Ideas*, that take their turns in our Understandings' (Locke, *Essay*, p. 182).
4. Hume, *Enquiry*, p. 19.
5. Hume, *Enquiry*, p. 58. Hume takes up this argument in another passage: '[i]t appears, then, that this idea of a necessary connexion among events arises from a number of similar instances, which occur, of the constant conjunction of these events; nor can that idea ever be suggested by any one of these instances, surveyed in all possible lights and positions' (Hume, *Enquiry*, p. 59). On this subject, see the third chapter of Malherbe, *La philosophie empiriste de David Hume*.
6. Hume, *Enquiry*, p. 58.
7. See the essay entitled 'Hume' in Deleuze, *Desert Island*.
8. See Bradley, *Appearance and Reality*. The polemical relationship between the pragmatists and the idealists took place principally in two philosophy journals at the beginning of the century: *Mind* and *Journal of Philosophy*. The most important interventions were from Schiller, Lovejoy and Dewey on the pragmatist side and from Bosanquet on the idealist side. Among the most memorable articles are: Schiller, 'Empiricism and the Abolute'; Schiller, 'On Preserving Appearances'; Lovejoy, 'The Thirteen Pragmatisms'; Bosanquet, 'Contradiction and Reality'. For an analysis of Anglo-American philosophy, at once historical and speculative, see Wahl, *Pluralist Philosophies of England and America*.
9. Whitehead, *Adventures of Ideas*, p. 204.
10. Whitehead, *Adventures of Ideas*, p. 280.
11. In this way, for instance, 'a molecule is a historic route of actual occasions; and such a route is an "event"' (PR, 80).
12. Butler, *Life and Habit*, p. 109.
13. In this stacking up of realities we find Leibniz's idea that '[e]ach

portion of matter may be conceived as like a garden full of plants and like a pond full of fishes. But each branch of every plant, each member of every animal, each drop of its liquid parts is also some such garden or pond' (Leibniz, *Monadology*, p. 256). And the analogy with Butler is all the more powerful if we think that, for Leibniz, 'each living body has a dominant entelechy, which in an animal is the soul; but the members of this living body are full of other living beings, plants, animals, each of which has also its dominant entelechy or soul' (*Monadology*, p. 257). The creation of these relations between multiplicity, the overlapping of scales and the soul or entelechy of each mode of existence makes Leibniz the originator of a movement that the theory of societies inherits, a theory present simultaneously in Tarde and Whitehead.

14. Butler, *Life and Habit*, p. 110.
15. Butler, *Life and Habit*, p. 124.
16. Stengers, *Thinking with Whitehead*, p. 45.
17. The biologist Pierre Sonigo has developed ideas that are surprisingly close to a non-reductionist theory of extensive phenomena. See Sonigo and Kupiec, *Ni Dieu ni gène*.
18. Wahl, *Vers le concret*, p. 157. (Translator's note: my own translation.)
19. Wahl, *Vers le concret*, p. 152. (Translator's note: my own translation.)

12

Nature and Societies

Order and Disorder

What is nature? So far, it has not been necessary to step outside of *Process and Reality*'s speculative project. Nature is fully deployed within the theory of existence and, in the theory of societies, it takes on the form of a singular type of empiricism. Doesn't this new question of nature, of what composes it, of its relations to existences, require, for the first time, stepping outside of speculative philosophy? Doesn't the question 'what is nature?' necessarily fall within a *philosophy of nature* of the kind that interests Whitehead in his earlier works and to which he returns in *Modes of Thought*? Several commentators on *Process and Reality* have embarked on this path by transporting, with differing degrees of obliqueness, the theory of actual entities on to a different plane from that of speculative philosophy. This approach, however, risks diminishing the *final* character of actual entities, reducing the importance and restricting the field of application of the ontological principle itself. Remember: there is nothing beyond actual entities; 'the rest is silence'. Neither a superior form of existence nor a nature *qua* principle of movement, becoming or creation can ever explain actual entities. The terms usually employed to characterise nature, however, tend to presuppose an ontological precedence of nature over the existences that compose it.

It should be said right away that the question of nature does not necessarily require a 'philosophy of nature', if one understands by the latter a philosophy in which nature would be a first and constitutive principle of all other forms of existence. The originality of *Process and Reality* is how it maintains the question of nature within a speculative approach in which nature is not *that which explains* but *that which has to be explained*. This question forms the object of one of the most important chapters in *Process and Reality*: 'The Order of Nature'. Its title already indicates how the problem is laid out: the question 'what is nature?' is connected to another question, namely, 'what is an order?' The fact that Whitehead puts these two concepts – order and nature – into direct relation is crucial.[1] He makes the question of order an essential function involved in the definition of speculative philosophy: '[t]hese chapters are concerned with the allied problems of "order in the universe", of "induction", and of "general truths"' (PR, 83). If it is difficult to respond to the question 'what is nature?' – since it requires a level of generality, a leap into a completely different register for which what has been developed so far would be of little use – it is, on the other hand, much easier to respond to the second question.

If Whitehead speaks of 'The Order of Nature' we should not be misled: '"[o]rder" is a mere generic term: there can only be some definite specific "order", not merely "order" in the vague' (PR, 83). Speaking of the order of nature in the singular is simply to place the accent on the *generic* characteristic of nature. The reality of order must, however, be located in diversity: local and specific orders, situated at this or that level. The tendency to reduce the multiplicity of orders in favour of a paradigmatic form of order in general[2] 'arises from the disastrous overmoralization of thought' (PR, 84). Unable to accept the heterogeneity of orders, thought projects a finality on to this heterogeneity, tracing it to a common form. Thought often produces order ready-made, locally distributed and realised once and for all. Whitehead, detecting a kind of projection, totally rejects the mode in which the problem of order is generally approached, a mode in which there exists a plurality of given solutions to the question of the relations between a general order and its specific forms. A reversal is effected. If the idea of general order emerges out of thought's overmoralisation, then refusing the latter entails taking into account a radical plurality of orders.

What are these 'orders' within nature? What does the idea refer to? Principally to societies, or at least to certain aspects of them. The reality of order is not found in a particular form of existence. It is neither an organising principle of things nor a finality to which they are drawn. It is simply one of the *consequences* of societies. This is why the approach to the question of order in *Process and Reality* is essentially *pragmatic*. In fact, if a society is indeed an enduring nexus, then the 'foundations' of order are found only in this *persistence*. So much so that, at this stage, the only definition we can give of order is: *everything that endures is an order, every order is a duration*. This is a minimal definition, but it shifts the question considerably by removing it from a rational principle that would pass through a succession of causes and effects, of principles and consequences, a conception based on the unearthing of foundations. This definition says nothing, of course, about the forms that an order might take, in the same way that the mere fact of a thing's persistence says nothing about its specificity or about what composes its duration. It merely implies that, on rare occasions, nexūs persist in the complex of disjunctive diversity, maintaining existential identity – this is what Whitehead calls 'order'. Order, in this sense, has an exceptional character that speculative philosophy needs to bring to light. Order takes on a precious, 'capital importance', the measure of which must be taken. We deal principally with orders in our perceptual experience, a fact that confirms its specificity once and for all.

Order, ultimately, is a relative term. Its only positivity comes from the notion of duration, with which it tends to be identified, though it emphasises a different dimension. It is, for the most part, what stands out from disorder.[3] One might think that distinguishing order and disorder brings little light to order's importance and, as such, to the search for a definition of nature. It is within this distinction, however, that the function of the concept of order should be situated, order being what is removed from disorder. Technically, disorder carries two separate meanings in *Process and Reality*. Concretely, however, these two meanings refer to one and the same thing. First: disorder is simply that which 'precedes' order. It is the group of realities that pre-exist a given order, at least from the point of view of that particular order in its process of being constituted. Pre-existence is fully relative to the fact that a new form of order emerges as soon as it is posed. It can, then, be said that '[a] society arises from disorder, where "disorder" is defined by reference to the ideal for that society' (PR, 91). According to the perspective of this ideal, everything that precedes is disorder, and the difference between the two – the new society and that which pre-exists it – is located on a moving line, in a continuous movement, across the society's history. The difference, then, is never clearly definable, it is never identified once and for all: just as a society depends on its inheritances, repetitions and transmissions, differentiation is reproduced at every stage, with the same risk that the society's identity may no longer maintain its 'own ideal'.

Secondly: disorder is what 'surrounds' the society, it is the society's environment, understood in a broad sense. Disorder, then, is what the society is not, what does not enter into its determining character. It is everything contemporary to the society that lacks a direct meaning for it, the society's neighbourhood extending to everything that exists: 'Beyond these societies there is disorder, where "disorder" is a relative term expressing the lack of importance possessed by the defining characteristics of the societies in question beyond their own bounds' (PR, 92). One might assume that this is a difference between the 'interior' and the 'exterior', the 'inside' and the 'outside', completing the difference between the 'anterior' and the 'actual'. It is nothing of the sort. The concepts of interiority and exteriority need to be refined, since zones of disorder are internal to societies as much as they are external, just as anteriority is contemporary with actuality. Disorder is what is not connected to the society's 'own ideal', and there exist a multitude of elements within every society that do not participate in it, do not matter for it. The same relativity applies, here: only from the perspective of a certain society, of a certain social organisation, from a *situated* point, can somewhere 'beyond' be understood as disorder.

Order and disorder, then, are relative terms. A common error consists in considering them independently of each other, reifying them as if they expressed particular realities with their own qualities and forms of existence. In both cases it is a question only of actual entities. They differ as soon as a society emerges that unfolds an identity that comes to contrast with the environment in which it is formed through repetition. But this environment, existing for a particular society, is not a pure disorder. It too can contain other societies, other orders. From a determined society's point of view, other societies can constitute disorder simply to the extent to which other organisational forms are not necessarily relevant to its own existence. What appears to be an order from one perspective, then, can be transformed into a disorder by changing the point of view. There is no possible position from which we could survey, perceive, analyse or think areas of order and their relations. This is due to the very nature of the concepts of order and disorder, not to any insufficiency in our modes of experience:

it is essential for the very existence of a society, as I will show, that it can consider a series of elements and domains of existence as disorder. What comes first are fluctuations of orders, arrangements between societies, an economy of stabilities and instabilities. All that can be said is that a society is constituted in a certain place and that it persists as long as it can according to the variations of its environment and according to areas of instability both external and internal to it. In this way, '[o]f course, the remote actualities of the background have their own specific characteristics of various types of social order. But such specific characteristics have become irrelevant for the society in question by reason of the inhibitions and attenuations introduced by discordance, that is to say, by disorder' (PR, 90).

Societies, Environments

We can now return to the question 'what is nature?' Whitehead does not seek a precise definition. He leaves the question open, limiting himself to the construction of the problem's set of terms. A definition, however, can be offered on the basis of particular elements connected to the theory of societies and the question of order. 'Nature' is *the structured society of all societies*. This definition has the form of a hypothesis: its only justification is that it maintains the coherence of the theory of societies, responding to the question of what nature is on the basis of such coherence. Two effects of this definition seem fundamental. First: it implies that nature is simply a technical term for expressing the society that, through its extension, is in some way ultimate. Every society that we experience, whatever its level of complexity – whatever its number of underlying nexūs – is, by definition, a part of nature. Nature, then, is not a space that contains things, but nor is it a substantial reality. It is, like all societies, a multiplicity of orders, of nexūs, each composed of becomings holding together across the same historic route. It is, then, an event that emerges from reciprocal prehensions between distinct orders, at once autonomous and connected by extensive relations. Secondly: this definition connects 'nature' to 'duration', but in a highly distinct way. The idea of nature as duration is hardly new; it can be found in most philosophies of nature. Here again, however, is Whitehead's opposition to philosophies that make a primary and continuous reality out of duration. Nature is a duration composed of the repeated inheritances operated by a multiplicity of orders and series. There is an 'adventure' of nature just as there is an adventure proper to each society.

It is impossible to think this adventure. Technically, we are absorbed in particular societies, positioned on a scale of experience. If we say that nature is the structured society of all societies then we also have to accept the relativity of our experience, the particular world relative to our 'power of acting'.[4] We are societies among others. Just as the organ requires the existence of the body only in a vague way, the contents of the body being outside of its reach, we experience nature as an extended existence to which we are linked. Certain connections are obvious since they correspond to our modes of experience, while others are more obscure, despite the fact they may change. Whitehead's position involves a refusal to translate nature into our modes of perception and experience. It could be said that in this specific way he joins together with Spinoza, who affirms that substance encompasses an infinite number of attributes, only two of which are accessible.[5]

Like structured societies in general, nature forms an environment for other societies. It should not be concluded, however, that every environment is a society – there can be non-social environments – though every structured society is an environment among others for the society subordinate to it. This environment might vary, of course, according to the relations maintained between its diverse components, leading to local modifications in other places. The environment of a given society can then be modified, raising the question of the society's stability and survival.

> The doctrine that every society requires a wider social environment leads to the distinction that a society may be more or less 'stabilized' in reference to certain sorts of changes in that environment. A society is 'stabilized' in reference to a species of change when it can persist through an environment whose relevant parts exhibit that sort of change. If the society would cease to persist through an environment with that sort of heterogeneity, then the society is in that respect 'unstable'. (PR, 100)

Stability is the capacity to persist across environmental variations, modifications of larger societies in which a given society is situated. In most cases these modifications lead to one of the following: the society's indifference, transformation or disappearance.

I will consider only the first two possibilities; but note, first, that a society never owes its disappearance, its death, to any internal reason, at least if 'internal' carries the sense, not of *interior*, but of a 'common element'. If a society emerges it prolongs its own mode of existence through the constraints belonging to its members. A society's disappearance is connected to the effects of an encounter: a modification of the environment no longer allows it to maintain its heritage, or the unstable elements that compose it no longer allow its heritage to be repeated. This is a complex relation and might be analysed in terms of power: does a society have the capacity to be indifferent to changes in the environment, or can it modify itself? This is the question that determines its survival. Societies have particular capacities, allowing them to respond to their encounters with heterogeneous environments in different ways. Certain societies are capable of change, others can guard their identity across sometimes momentous environmental changes, and there are others still that are unable to tolerate even the slightest environmental shift.

To understand these differences, these forms of negotiation between societies and environmental variations, two clarifications must be made concerning social organisation. First, a society is 'complex' to greater and lesser degrees. As with the other categories mobilised here, 'complexity' is fundamentally relative; it simply indicates the importance of a society's extension. Instead of referring to a particular society's nature, complexity refers to *empirical*, or 'quantitative' differences between societies: 'A "structured society" may be more or less "complex" in respect to the multiplicity of its associated sub-societies and sub-nexūs and to the intricacy of their structural pattern' (PR, 100). A society, then, can be more or less 'specialised'. Specialisation is the capacity to maintain oneself in a specific environment. The notion 'seems to include both that of "complexity" and that of strictly conditioned "stability"' (PR, 100). If a specialised society is stable, it is simply because it has an environment adequate to its form of organisation. Any variation in this environment could pose a threat to its survival. Certain highly specialised societies can survive only within a 'very specialized sort of environment' (PR, 101).

These two concepts, then – complexity and specialisation – are consequences of the way societies are organised, the relations between their members. Certain societies include an ample multiplicity of sub-societies, while others contain only a few. Some of these sub-societies have extremely diverse characteristics, producing a more flexible level of specialisation, while others are held in an existential identity that denies their capacity to maintain themselves outside of specific environments. Organisations determine aptitudes, the particular powers that societies have vis-à-vis their environments.

Stability and Innovation: Life in Nature

The problem 'for Nature is the production of societies which are "structured" with a high "complexity", and which are at the same time "unspecialized". In this way, intensity is mated with survival' (PR, 101). A society's survival depends on its aptitudes relative to variation. As shown above, there are two possible ways of responding to an environmental transformation: an *indifference* to environment or a *metamorphosis* of the society itself. These are the cornerstones of two distinct regimes of existence: physical bodies and living societies. 'Physical' societies are defined 'by eliciting a massive average objectification', made possible 'by eliminating the detailed diversities of the various members of the nexus in question' (PR, 101). Essentially, then, they function through averages, to the extent that we can say that every physical object owes its survival to its internal average, its disregard of details. Such an object overwhelms 'the nexus by means of some congenial uniformity which pervades it. The environment may then change indefinitely so far as concerns the ignored details – so long as they can be ignored' (PR, 101). Physical societies persist as much as they can across variations, that is, as much as it is possible to ignore the effects of changes upon their own structure, within the sub-nexūs that compose them.

> These material bodies belong to the lowest grade of structured societies which are obvious to our gross apprehensions. They comprise societies of various types of complexity – crystals, rocks, planets, and suns. Such bodies are easily the most long-lived of the structured societies known to us, capable of being traced through their individual life-histories. (PR, 102)

The reason for their survival has to do with their *power of averaging*, their elimination of originality. This power makes them one of the most durable regimes in structured societies. Averages are some of the most effective conditions for the survival of stability.

But there is another possible answer that, instead of passing through averages, touches on invention, 'an initiative in conceptual prehensions, i.e., in appetition' (PR, 102). Remember that conceptual prehensions are prehensions of eternal objects that determine the manner – the how – in which data is captured and integrated into actual entities. In general, however, '[a] society does not in any sense create the complex of eternal objects which constitutes its defining characteristic. It only elicits that complex into importance for its members, and secures the reproduction of its membership' (PR, 92). This is why a society is essentially a *repetition*, a *tradition*. It repeats an ancestral lineage imposed on its every member. The originality of this response to environment concerns a new capacity, another power that certain societies have: *they take the conceptual initiative*. In other words, in certain circumstances, societies are capable of modifying their organisation, introducing *innovation* into their orders. Everything happens as if certain societies could produce, simply out of themselves, distinct forms of relation to their environments, diverse economies of relations and contrasts between their subaltern societies. Whitehead calls this facility 'life'.

Life, then, 'is the name for originality, and not for tradition' (PR, 104). Although Whitehead speaks of 'living societies', life as such cannot be identified with any specific society. The reasons for this impossibility are clear. Not only would such an identification challenge what makes life distinct, namely, that it is an 'innovation' and not a repetition, but more importantly it would go back to considering life as part of a particular domain, in the way we speak of the 'living' as distinct from the 'inert'. There is no strict boundary, however, between the 'physical' and the 'living', 'there is no absolute gap' (PR, 102). It is simply that, for certain societies, the slightest life can be essential, while for others it can be merely academic. This is why Whitehead writes that '[l]ife lurks in the interstices of each living cell, and in the interstices of the brain' (PR, 105–6). Life is interstitial, it is an innovation within order, in the interstices of the order's repetitions and reproductions. 'Living societies' are simply those for whom these interstitial zones have taken on the most importance. As Stengers writes: 'the singularity of living societies, what justifies them as such, should be called a "culture of interstices"'.[6] No society can be identified with its interstices, and yet the significance of the latter, their inclusion, is what distinguishes living organisations from others. This is why *Process and Reality* has no theory of life, affirms no vital principle. This would amount to attempting to think novelty as novelty, as abstract originality. But there is only one way to account for life: to introduce it into repetition, into habit and order, and then to experience the areas where something new is produced.

'Life', then, intensifies the fact that there can be knowledge only of the singular – of *this* society, *this* duration, *this* novelty. This knowledge is singular not only because it has to do principally with societal 'adventures', but also because knowledge itself is involved in societies. Captured in the trials of what a society can do, knowledge has no a priori guarantee of success.

> The Castle Rock at Edinburgh exists from moment to moment, and from century to century, by reason of the decision effected by its own historic route of antecedent occasions. And if, in some vast upheaval of nature, it were shattered into fragments, that convulsion would still be conditioned by the fact that it was the destruction of that rock. The point to be emphasized is the insistent particularity of things experienced and of the act of experiencing. Bradley's doctrine – Wolf-eating-Lamb as a universal qualifying the absolute – is a travesty of the evidence. That wolf eats that lamb at that spot at that time: the wolf knew it; the lamb knew it; and the carrion birds knew it. (PR, 43)

Notes

1. 'For the organic doctrine the problem of order assumes primary importance' (PR, 83).
2. Leibniz, in the *Discourse on Metaphysics*, poses the relations between general and specific order: '[b]ut it is well to bear in mind that God does nothing out of order. So, whatever passes extraordinary is only so in relation to some particular order established among creatures. For, in relation to the universal order, everything conforms to it. So true is this, that not only does nothing happen in the world that is absolutely irregular, but such a thing cannot even be imagined' (*Discourse on Metaphysics*, p. 44). In this passage we find the 'extraordinary' character that appears when we consider only a particular order, and the disappearance of this character when we place it directly on a level of 'universal order'. Whitehead puts himself, then, into an opposing economy to that of Leibniz.

3. By posing the question in terms of an *emergence* from disorder, Whitehead revives the question of the *production of order* found particularly in Hume's empiricism. Of course, the differences are important, and the same reservations emphasised above concerning other subjects still apply. The fact remains, however, that the problem's heritage is to be found in terms posed by Hume. On this subject, see the key chapter that Malherbe has dedicated to this question, entitled 'Genèse de l'entendement', in *La philosophie empiriste de David Hume*.
4. See von Uexküll, *A Foray into the Worlds of Animals and Humans*.
5. See Guéroult, *Spinoza*, p. 52.
6. Stengers, *Thinking with Whitehead*, p. 328.

13

Conclusion: What is Speculative Empiricism?

Selection and Speculation

Process and Reality ends with a warning: '[t]he chief danger to philosophy is narrowness in the selection of evidence' (PR, 337). Although this danger of narrowness might emerge from the 'idiosyncrasies and timidities of particular authors, of particular social groups, of particular schools of thought, of particular epochs in the history of civilization' (PR, 337), we should not be mistaken: it occurs within philosophy, in its activity, its method. And the fact that this issue arises at the end of *Process and Reality* reveals the ambition that has accompanied its composition: Whitehead has resisted this danger through the form and ambition of his speculative construction. The temptation of a narrowness in selection attempts to expel speculative philosophy at the same time as it haunts each part of its system.

At the origin of the 'selection' is a confusion between the methods of science and the methods of philosophy. The former have different modes and aims. 'The field of a special science is confined to one genus of facts, in the sense that no statements are made respecting facts which lie outside that genus' (PR, 9). The knowledge formed within this 'field' is termed a 'specific knowledge', and is possible only by way of a set of operations of simplification and selection that inscribe 'facts' into a space of reference. Beneath the seeming overtness of the statements of a specific knowledge, radical transformations of experience take place, the most obvious expression of which is the 'bifurcation of nature' that characterises the modern sciences.

The method proper to speculative philosophy is expressed in a general maxim in the form of an imperative: '[p]hilosophy can exclude nothing'.[1] This is both an evaluative principle of systems

(What must it reject? Which part of experience is withdrawn?) and a structural requirement. It seems as if selection is at the heart of the activity of thought and 'specific knowledge' an extension of that activity. To think is often to abstract. As such, then, another function is attributed to philosophy: *to invert the habitual course of thought*. The habitual movement is reversed: it no longer moves from experience towards simplification but rather from simplification towards experience. Philosophy reminds us of elements and general principles covered over by the habits of thought. Its ambition is to 'attempt to make manifest the fundamental evidence as to the nature of things'[2] as well as to express the particular routes that experience can take and, therefore, the selections it can make.

Every system has its limits. The insufficiency of intuition and language when associated with the 'creative advance of the universe', as I have shown, guarantees the failure of every project that offers itself up as a final answer. It does not necessarily follow, however, that there are elements in experience that are unknowable by nature.

> Thus a complete understanding is a perfect grasp of the universe in its totality. We are finite beings; and such a grasp is denied us.
>
> This is not to say that there are finite aspects of things which are intrinsically incapable of entering into human knowledge. [. . .] we can know anything in some of its perspectives. But the totality of perspectives involves an infinitude beyond human knowledge.[3]

The finitude of thought is not accidental, and philosophy inherits it. But maintaining the requirement to which Whitehead connects the very origin of philosophy is essential: the attempt to elucidate ultimate principles mobilised by the entirety of experience. This is a tension spanning *Process and Reality*: 'the difficulty of philosophy lies in the two opposing conditions it has to satisfy: 1) it can depart only from immediate experience; 2) it aims to extricate the ultimate principles of experience as a whole'.[4] If philosophy takes immediate experience as its point of departure, and if it aims at nothing more than the interpretation of such experience, it remains, nevertheless, a search for ultimate principles.

The Components of a Philosophy of Experience

In this book I have attempted to approach Whitehead's work as a resistance to a philosophy of experience that would impose a priori limits on experience, that would determine its space. If the concept of experience is 'one of the most deceitful in philosophy',[5] it is because it is always limited, constructed inside of other problems – those of perception, intension or consciousness – as if we could look elsewhere for its conditions and justification. A philosophy that would treat experience *as such* is still to be constructed and continually to be remade. My ambition, here, has been to explore some of its components in the form of what I have proposed to call a 'speculative empiricism'.[6]

It is, before all else, a *technical* philosophy, the requirement of which – to exclude nothing from experience – passes through the creation of a set of abstractions, ideas that have no function outside of demonstrating characteristics of experience that, without them, would be at risk of elimination. At the same time as it is an empiricism, then, it is a radical rationalism. It fully accepts the conditions of coherence, logic and necessity; it is a pure treatment of ideas. In *The Fold*, Deleuze sees Whitehead's work as one of the last great systematic metaphysics, in the tradition of a philosophical form introduced by Leibniz.

> He [Whitehead] takes up the radical critique of the attributive scheme, the great play of principles, the multiplications of categories, the conciliation of the universal and the individual example, and the transformation of the concept into a subject: an entire hubris.[7]

Whitehead's empirical and speculative requirements could not be better described: every concept becomes a subject, every universal a particular case and vice versa. The pyramid, the soldiers, the lightning strike – these singular cases become concepts, crisscrossed by a multiplicity of abstractions that have themselves become universals since they are present in every part of experience. And actual entities, eternal objects – purely abstract concepts – are transformed into subjects because, once invented, they are added to the elements of immediate experience. All of these examples escape evocation and metaphor to literally become concepts. And, in a reciprocal movement, the concepts are stripped of all abstraction to become components of experience. Experience is not the same after the invention and consideration of the abstractions produced by speculative philosophy, as is the case for the introduction of any idea in general. This is the first condition of a philosophy of experience: concepts are singular and universal, but so is every element of experience if thought adequately.

Speculative empiricism, however, is not limited to seeking a relation between concepts and experience with a method of interpretation. It tries to offer an explanation of immediate experience and is unprecedented in its ambition and form. I have analysed this aspect at length through the theory of societies. My guiding hypothesis was that the theory of societies distributed in *Process and Reality* is at the centre of this project, and forms the second great condition of a speculative empiricism. It is the theory of societies that genuinely connects the abstraction of actual entities with immediate experience. It is the real empirical test of the system, its point of 'ultimate control': 'the more general the rationalistic scheme, the more important is this final appeal [to experience]' (PR, 17).

To understand this, and to give full meaning to the empirical dimension of speculative philosophy, we can cite a passage from Whitehead's *Symbolism*:

> the conception of the world here adopted is that of functional activity. By this I mean that every actual thing is something by reason of its activity; whereby its nature consists in its relevance to other things, and its individuality consists in its synthesis of other things so far as they are relevant to it.
>
> In enquiring about any one individual we must ask how other individuals enter 'objectively' into the unity of its own experience.[8]

Whitehead says it very clearly: this is a 'conception of the world'. This conception makes more intuitive, it *makes seen* and *makes felt* on the level of experience, what *Process and Reality* attempts to think speculatively. Experience is not made of things, of individuals whose identities would be fixed and determined once and for all and whose boundaries would be tightly delimited. It is made of 'functional activities'. The parts of experience pass into each other, they transform each other through reciprocal actions and mutual influences. As such, '[a] flower turns to the light with much greater certainty than does a human being, and a stone conforms to the conditions set by its external environment with much greater certainty than does a flower'.[9] There is no part of experience that is not crossed and not constituted by actions coming from other areas. So much so that, step by step, everything is submitted to the influence of everything else.

> We find ourselves in a buzzing world, amid a democracy of fellow creatures; whereas, under some disguise or other, orthodox philosophy can only introduce us to solitary substances, each enjoying an illusory experience. (PR, 50)

Whitehead also calls these 'functional activities' 'events'. The concept of the event becomes generic and is applied to every element of an experience: '[a]n event does not just mean that "a man has been run over". The Great Pyramid is an event, and its duration for a period of one hour, thirty minutes, five minutes . . . a passage of Nature, of God, or a view of God.'[10] This concept replaces every category that attempted to express the individuality of factors of experience.

This 'conception' is clear when it remains at this scale. But as soon as we try to clarify its terms, countless difficulties emerge. How, for instance, can the presence of 'other things' within the individuality of one of those things be explained? How do we maintain that process and individuality require each other[11] without falling into a generality of principles? If selection is indeed a danger peculiar to philosophy, metaphor poses just as much danger, since it dispenses with what gives it specificity, namely, its capacity to rationally account for experience. From the moment we occupy this level of explanation, however, as soon as we take events as the departure point for a speculative analysis, we can only give these questions meaning in vague and evocative forms. This is why – and this point is fundamental – *events are not what explain but what must be explained.*

Most philosophies of experience – empiricism in particular – confuse the order of reasons. They assume that the order in which problems are demonstrated should be the same as the order of experience. They take what seems obvious in experience (sensations, perceptions, events) as the initial point of philosophical explanation, when they should be its final one. *Process and Reality* totally overturns this. The qualities of events – duration, extension and change – rather than being the foundation of speculative philosophy, must, at every stage, be replaced by a constellation of concepts that makes possible their conceptual expression. Duration, then, is explained by the epochal theory of becoming, as rhythm, inheritance and repetition. Extension is accounted for by existing relations between groupings of actual entities. And, finally, change is explained by the theory of orders and by instability. One more time, since this point is crucial: if speculative empiricism is a philosophy of events, it is not one in the sense that it uses events to begin its explanations. It is a philosophy of events because it attempts to explain events, or, more precisely, because it brings to light what they require. Speculative empiricism is indeed a radical empiricism, but it is based on methods alien to empiricism, techniques of a whole other order: the invention of abstractions and techniques for interpreting experience. Its goal is the 'elucidation of immediate experience' (PR, 4).

> The use of philosophy is to maintain an active novelty of fundamental ideas illuminating the social system. It reverses the slow descent of accepted thought towards the inactive commonplace. If you like to phrase it so, philosophy is mystical. For mysticism is direct insight into depths as yet unspoken. But the purpose of philosophy is to rationalize mysticism: not by explaining it away, but by the introduction of novel verbal characterizations, rationally coordinated.[12]

Notes

1. Whitehead, *Modes of Thought*, p. 2.
2. Whitehead, *Modes of Thought*, p. 48.
3. Whitehead, *Modes of Thought*, p. 42.
4. Saint-Sernin, *Whitehead, un universe en essai*, p. 34. (Translator's note: my own translation.)
5. Whitehead, *Symbolism*, p. 16.
6. My commitment to redefining empiricism has required me to leave certain key questions in *Process and Reality* unresolved, in particular the theory of propositions and the function of God. These form the central axes of Stengers's *Thinking with Whitehead.*
7. Deleuze, *The Fold*, p. 76.
8. Whitehead, *Symbolism*, p. 26.
9. Whitehead, *Symbolism*, p. 42.
10. Deleuze, *The Fold*, p. 76.
11. 'Process and individuality require each other. In separation all meaning evaporates. The forms of process (or, in other words, the

appetition) derive their character from the individual involved, and the characters of the individual can only be understood in terms of the process in which they are implicated' (Whitehead, *Modes of Thought*, pp. 96–7).

12. Whitehead, *Modes of Thought*, p. 174.

Works Cited

Works by Alfred North Whitehead

The Concept of Nature (Cambridge: Cambridge University Press, 1920).

Science and the Modern World (New York: Pelican Mentor, 1948) (first published 1925).

Symbolism: Its Meaning and Effect (Cambridge: Cambridge University Press, 1928).

Process and Reality: An Essay in Cosmology, ed. David Ray Griffin and Donald W. Sherburne (New York: The Free Press, 1978) (first published 1929).

Procès et réalité. Essai de cosmologie, trans. Daniel Charles, Maurice Élie, Michel Fuchs, Jean-Luc Gautero, Dominique Janicaud, Robert Sasso and Arnaud Villani (Paris: Gallimard, 1995).

Adventures of Ideas (New York: The Free Press, 1961) (first published 1933).

Modes of Thought (New York: The Free Press, 1968) (first published 1938).

Essays in Science and Philosophy (New York: Philosophical Library, 1947).

Works on Alfred North Whitehead

Cesselin, Félix, 'La Bifurcation de la Nature', *Revue de Métaphysique et de Morale* 1 (1950).

—*La philosophie organique de Whitehead* (Paris: Presses Universitaires de France, 1950).

Christian, William, *An Interpretation of Whitehead's Metaphysics* (New Haven, CT: Yale University Press, 1959).

Dumoncel, Jean-Claude, *Les sept mots de Whitehead ou L'aventure de l'être: créativité, processus, événement, objet, organisme, 'enjoyment',*

aventure: une explication de 'Processus et réalité (Paris: EPEL, 1998).
Élie, Maurice, 'Sur l'*Ultime*. A propos d'une catégorie de *Procès et Réalité*', *Les Études Philosophiques* 63.4 (2002).
Jones, Judith, *Intensity: An Essay in Whiteheadian Ontology* (Nashville, TN: Vanderbilt University Press, 1998).
Leclerc, Ivor, *Whitehead's Metaphysics: An Introductory Exposition* (London: Allen and Unwin, 1958).
Saint-Sernin, Bertrand, *Whitehead, un universe en essai* (Paris: Vris, 2000).
Stengers, Isabelle, *Penser avec Whitehead. Une libre et sauvage création de concepts* (Paris: Seuil, 2002).
—*Thinking with Whitehead*, trans. Michael Chase (Cambridge, MA: Harvard University Press, 2011).
Wahl, Jean, *Vers le concret: Études d'histoire de la philosophie contemporaine* (Paris: Vrin, 1932).

General Works

Aristotle, 'Categories', in *The Complete Works of Aristotle*, ed. Jonathan Barnes (Princeton: Princeton University Press, 1984).
—'Metaphysics', in *The Complete Works of Aristotle*, ed. Jonathan Barnes (Princeton: Princeton University Press, 1984).
—'Physics', in *The Complete Works of Aristotle*, ed. Jonathan Barnes (Princeton: Princeton University Press, 1984).
Barthes, Roland, *Michelet*, trans. Richard Howard (Berkeley: University of California Press, 1987).
Bergson, Henri, *The Creative Mind: An Introduction to Metaphysics*, trans. Mabelle L. Andison (Mineola, NY: Dover Publications, 2007).
—*Key Writings*, ed. Keith Ansell Pearson and John Mullarkey (New York: Continuum, 2002).
—*Matter and Memory*, trans. Nancy Margaret Paul and W. Scott Palmer (New York: Zone Books, 1988).
Bosanquet, Bernard, 'Contradiction and Reality', *Mind* 15.57 (1906).
Bradley, F. H., *Appearance and Reality: A Metaphysical Essay* (London: Allen, 1908).
Brahami, Frédéric, *Introduction au Traité de la nature humaine de David Hume* (Paris: Presses Universitaires de France, 2003).
Bréhier, Émile, *Schelling* (Paris: Félix Alcan, 1912).
Butler, Samuel, *Life and Habit* (London: A. C. Fifield, 1910).
Caye, Pierre, 'Destruction de la métaphysique et accomplissement de

l'homme (Heidegger et Nietzsche)', in *Heidegger et la question de l'humanisme*, ed. Bruno Pichard (Paris: Presses Universitaires de France, 2005).

Combes, Muriel, *Simondon: individu et collectivité: pour une philosophie du transindividuel* (Paris: Presses Universitaires de France, 1999).

De Freiberg, Dietrich, 'De ente et essentia', in Thomas D'Acquin and Dietrich de Freiberg, *L'etre et l'essence. Le vocabulaire médiéval de l'ontologie*, trans. and ed. Alain de Libera and Cyrille Michon (Paris: Seuil, 1996).

Deleuze, Gilles, *Desert Island and Other Texts, 1953–1974*, trans. Mike Taormina, ed. David Lapoujade (Los Angeles: Semiotext(e), 2004).

—*Difference and Repetition*, trans. Paul Patton (London and New York: Continuum, 1997).

—*The Fold: Leibniz and the Baroque*, trans. Tom Conley (London: The Athlone Press, 1993).

Deleuze, Gilles, and Guattari, Félix, *A Thousand Plateaus: Capitalism and Schizophrenia*, trans. Brian Massumi (Minneapolis: University of Minnesota Press, 1987).

—*What is Philosophy?*, trans. Hugh Tomlinson and Graham Burchell (New York: Columbia University Press, 1994).

Deleuze, Gilles, and Parnet, Claire, *Dialogues*, trans. Hugh Tomlinson and Barbara Habberjam (New York: Columbia University Press, 1987).

Descartes, René, 'Meditations', in *Discourse on Method and Meditations*, trans. Elizabeth S. Haldane and G. R. T. Ross (Mineola, NY: Dover Publications, 2003).

—'Meditations on First Philosophy', in *The Philosophical Writings of Descartes*, trans. John Cottingham, Robert Stoothoff and Dugald Murdoch (Cambridge: Cambridge University Press, 1985), vol. 2.

—'Principles of Philosophy', in *The Philosophical Writings of Descartes*, trans. John Cottingham, Robert Stoothoff and Dugald Murdoch (Cambridge: Cambridge University Press, 1985), vol. 1.

Eisendrath, Craig, *The Unifying Moment: The Psychological Philosophy of William James and Alfred North Whitehead* (Cambridge, MA: Harvard University Press, 1971).

Gilson, Étienne, *Jean Duns Scot* (Paris: Vrin, 1952).

Guéroult, Martial, *Spinoza* (Paris: Aubier-Montaigne, 1968), vol. 2.

Heidegger, Martin, *Aristotle's Metaphysics Θ 1–3: On the Essence and Actuality of Force*, trans. Walter Brogan and Peter Warnek (Bloomington: Indiana University Press, 1995).

—*Kant and the Problem of Metaphysics*, trans. Richard Taft (Bloomington: Indiana University Press, 1997).
Hume, David, *An Enquiry Concerning Human Understanding*, ed. Tom L. Beauchamp (Oxford: Clarendon Press, 2000).
—*A Treatise of Human Nature*, ed. David Faye Norton and Mary J. Norton (Oxford: Oxford University Press, 2000).
James, William, *Essays in Radical Empiricism*, ed. Fredson Bowers and Ignas K. Skrupskelis (Cambridge, MA: Harvard University Press, 1976).
—*A Pluralistic Universe*, ed. Fredson Bowers (Cambridge, MA: Harvard University Press, 1977).
—*Pragmatism*, ed. Fredson Bowers (Cambridge, MA: Harvard University Press, 1975).
—*The Principles of Psychology* (New York: Dover Publications, 1950), vol. 1.
—*Some Problems of Philosophy*, ed. Frederik Burkhardt, Fredson Bowers and Ignas K. Skrupskelis (Cambridge, MA: Harvard University Press, 1979).
—*The Will to Believe*, ed. Fredson Bowers, Frederick Burkhardt and Ignas K. Skrupskelis (Cambridge, MA: Harvard University Press, 1979).
Jullien, François, *Procès ou creation. Une introduction à la pensée chinoise* (Paris: Le livre de poche, 1989).
Lapoujade, David, *William James. Empirisme et pragmatisme* (Paris: Les empêcheurs de penser en rond, 2007).
Lautman, Albert, *Essai sur l'unité des mathématiques et divers écrits* (Paris: 10/18, 1976).
Leibniz, G. W., *Discourse on Metaphysics*, ed. and trans. R. N. D. Martin and Stuart Brown (Manchester: Manchester University Press, 1988).
—*The Monadology and Other Philosophical Writings*, trans. Robert Latta (Oxford: Oxford University Press, 1971).
—*New Essays on Human Understanding*, trans. and ed. Peter Remnant and Jonathan Bennett (Cambridge: Cambridge University Press, 1996).
Locke, John, *An Essay Concerning Human Understanding*, ed. Peter H. Nidditch (Oxford: Clarendon Press, 1975).
Lovejoy, A. O., 'The Thirteen Pragmatisms', *The Journal of Philosophy, Psychology and Scientific Methods* 5.1 (1908).
Malherbe, Michel, *La philosophie empiriste de David Hume* (Paris: Vrin, 1976).
Morgan, Lloyd, *Emergent Evolution* (London: Williams and Norgate, 1927).

Michaud, Yves, *Locke* (Paris: Presses Universitaires de France, 1998).

Newton, Isaac, *The Mathematical Principles of Natural Philosophy*, trans. Andrew Motte (New York: Daniel Adee, 1848).

Peirce, Charles Sanders, 'The Doctrine of Necessity Examined', in *The Collected Papers of Charles Sanders Peirce*, vol. 6, ed. Charles Hartshorne and Paul Weiss (Cambridge, MA: Harvard University Press, 1935).

—'How to Make Our Ideas Clear', in *The Collected Papers of Charles Sanders Peirce*, vol. 5, ed. Charles Hartshorne and Paul Weiss (Cambridge, MA: Harvard University Press, 1934).

—'On Signs and the Categories', in *The Collected Papers of Charles Sanders Peirce*, ed. Arthur W. Burks, vol. 8 (Cambridge, MA: Harvard University Press, 1958).

—'A Survey of Pragmatism', in *The Collected Papers of Charles Sanders Peirce*, ed. Charles Hartshorne and Paul Weiss, vol. 5 (Cambridge, MA: Harvard University Press, 1934).

Plato, 'Timaeus', trans. Donald J. Zeyl, in *Complete Works*, ed. John M. Cooper (Indianapolis: Hackett, 1997).

Pradeau, Jean-François (ed.), *Platon. Les forms intelligibles* (Paris: Presses Universitaires de France, 2001).

Ruyer, Raymond, *La genèse des forms vivantes* (Paris: Flammarion, 1958).

—*La gnose de Princeton* (Paris: Fayard, 1977).

Sauvanet, Pierre, *Le rythme grec d'Héraclite à Aristote* (Paris: Presses Universitaires de France, 1999).

Schelling, F. W. J., *First Outline of a System of the Philosophy of Nature*, trans. Keith R. Peterson (Albany, NY: State University of New York Press, 2004).

Schiller, F. C. S., 'Empiricism and the Absolute', *Mind* 14.55 (1905).

—'On Preserving Appearances', *Mind* 12.47 (1903).

Scotus, Duns, *Le Principe d'individuation* (Paris: Vrin, 1992).

Simondon, Gilbert, *L'individuation psychique et collective* (Paris: Aubier, 1989).

Sonigo, Pierre, and Kupiec, Jean-Jacques, *Ni Dieu ni gene* (Paris: Pocket/ Seuil, 2003).

Tarde, Gabriel, *Monadology and Sociology*, trans. Theo Lorenc (Melbourne: re.press, 2012).

—*L'Opposition universelle* (Paris: Les empêcheurs de penser en rond, 1999).

von Uexküll, Jakob, *A Foray into the Worlds of Animals and Humans*, *with A Theory of Meaning*, trans. Joseph D. O'Neil (Minneapolis: University of Minnesota Press, 2010).

Wahl, Jean, *Du rôle de l'idée de l'instant dans la philosophie de Descartes* (Paris: Vrin, 1953).

—*Pluralist Philosophies of England and America*, trans. Fred Rothwell (London: Open Court, 1925).

Worms, Frédéric, 'James et Bergson: Lectures croisées', *Philosophie* 64 (2000).

Index